Homo Faber and Homo Economicus in the Scientific Revolution

This book tells the story of how the "servile arts" turned into the "mechanical arts," which in turn developed into a kind of philosophical apparatus that made modern science possible.

Why did the scientific revolution take place in the West and not in China or the Islamic world? How did humanity's progress in science and technology, which had been moving along at a relatively steady pace for tens of thousands of years, end up taking such an unprecedented leap? Subjecting the history of thought and technology to a novel interpretation based on the relationship between theory and practice, Ahmet Selami Çalışkan argues that the industrial revolution and modern science—and the scientific revolution that preceded both—did not alone suffice to sort out the philosophical problems of their day or to produce the institutions of the modern age. Both required a new sort of human: *Homo economicus faber.*

Tracing the historical emergence of this figure and its persistence in our own age, this book offers an innovative and holistic assessment of the economic, cultural and political effects of centuries of interaction between East and West and their repercussions in our world today.

Ahmet Selami Çalışkan holds a master's degree in political history and international relations from Marmara University, where he also earned a PhD with his dissertation "The Paradigm and Founder Effect of Practice in Modern Science." He is the CEO of Tekhnelogos Software, an R&D center that conducts artificial intelligence–based projects. He is also the chairman of the board of directors of Istanbul Jazari Museum.

Homo Faber and Homo Economicus in the Scientific Revolution

Ahmet Selami Çalışkan
Translated by Zahit Atçıl

Routledge
Taylor & Francis Group

NEW YORK AND LONDON

First published 2022
by Routledge
605 Third Avenue, New York, NY 10158

and by Routledge
4 Park Square, Milton Park, Abingdon, Oxon, OX14 4RN

Routledge is an imprint of the Taylor & Francis Group, an informa business

© 2022 Ahmet Selami Çalışkan

Library of Congress Cataloging-in-Publication Data
A catalog record for this title has been requested

ISBN: 978-1-032-23107-5 (hbk)
ISBN: 978-1-032-23113-6 (pbk)
ISBN: 978-1-003-27575-6 (ebk)

DOI: 10.4324/9781003275756

Typeset in Times New Roman
by codeMantra

Contents

Figures

Preface

One of the questions that had remained in my mind for a long time since our studies on the philosophy of nature we began in 2002 at the Foundation for Science and Arts was the following: Why did/could mathematical or experimental science not develop or remain limited in Islamic thought in general and in Islamic theology *(kalam)* in particular? Since this question can be posed otherwise as, "Why did the Scientific Revolution occur in the West or was the reality actually that so?" one of the important questions occupying my mind was on the origins of epistemological changes and the process of turning them into (transformation) practice. I owe particular thanks to Prof. Dr. Ihsan Fazlioglu and Ass. Prof. Ishak Arslan, who were regulars of the foundation, for their encouraging me to study this subject. I am grateful for my advisor Prof. Dr. Ali Durusoy for his guiding me to transform the question in my mind into an academic work. I am also thankful to A.C. Crombie for a map of history of science and a pattern of the Latin counterparts.

I must first mention my father who encouraged me to study first engineering, then social science, saying, "Students of engineering may one day study social sciences but social scientists do not have the same chance," and my mother who never leaves me without prayers and my wife Sevda, who is an invisible supporter of all my studies.

Lastly, I thank George Mason University Ali Vural Ak Center for Global Islamic Studies who provided me the opportunity to write most part of this study.

'Tis written: "In the Beginning was the *Word*."
Here am I balked: who, now can help afford?
The *Word*?—impossible so high to rate it;
And otherwise must I translate it.
If by the Spirit I am truly taught.
Then thus: "In the Beginning was the *Thought*"
This first line let me weigh completely,
Lest my impatient pen proceed too fleetly.
Is it the *Thought* which works, creates, indeed?
"In the Beginning was the *Power*," I read.
Yet, as I write, a warning is suggested,
That I the sense may not have fairly tested.
The Spirit aids me: now I see the light!
"In the Beginning was the *Act*," I write.
Goethe, *Faust*

"No act, no knowledge; no knowledge, no act."
Ibn Kutayba, *Uyūn al-akhbār*

"Every knowledge not experienced (applied) is hanged on a point between true and false."
Al-Jazari, *Kitāb al Hiyal*

Introduction

Is it possible to read intellectual history in respect to the relationship between theory and practice or how can this relationship be interpreted in respect to intellectual history? The answer to this question is significant for how the concept of "science" as a basic element of the modern paradigm is determinative in both intellectual world and practical life. In fact, it would not be wrong to define the whole human adventure as something building a bridge between "to know" and "to make," because the relationship between the two is essential for conceiving and creating a "life."

Considering the phases that the modern epistemology went through especially in the last century, it seems to have maintained its integrity and continuity, and even its existence by being under siege by an industrial paradigm. Is this dependency relationship between epistemology and practical space a process built by itself as historical necessity and its natural *entelekhia* or can we talk about contingencies for the humanity or particularly for the West in various bifurcations? All philosophical inquiries on the modern science and even on the modernity also include secondary questions that require encountering, albeit indirectly, this question in the background.

In fact, Kant lists the questions of Enlightenment philosophy as, "What can we know?" "What ought we to do?" "What may we hope for?" and then asks the fourth question, "What is man?"[1] Even if he argues that these questions belong to metaphysics, philosophy, religion and anthropology respectively, modern science answers or dares to answer all these questions. The venture of modern science brings about the answer to how it encircles these questions and tells that the relationship between "to know" and "to make" sets an ontological ground for the human being to define himself.

In addition to the theoretical aspects of the contemporary philosophy, "practical aspects" that were reflected in almost all fields from

DOI: 10.4324/9781003275756-1

ethics to linguistics are discussed in various platforms through the results of the practical or their epistemological connections. In its widest expression, the idea that the path leading to the industrial revolution started necessarily by the scientific revolution[2] and the mathematization of the philosophy of nature and the determination of its laws were the beginning of this process. This idea is necessarily based on the assumption that everything that appeared in the West was unique and the perception of nature and science before then inhibited the "progress" or did not allow the diversification of material culture.

Most of the appraisals on the historical background of the modern science usually maintain that the deficiency of the classical *epistheme*, the long-lasting domination of the Aristotelian system, the obstruction of the scholastic thought to "think," the insufficiency of cosmological definitions —perhaps naturally—after the collapse of all these in time. All these appraisals, however, remain to be only descriptive in nature and fail to expose what kind of knowledge filled what kind of space, why people sought a new kind of knowledge and what kind of ontology these accumulations created. This book works methodologically on the connection between the individual mentalities and institutional transformations and the formation of spaces. One of these factors only, for example, the intellectual ventures of philosophers or the emergence of institutions, would fall into the field of analyzing a sophisticated and hypothetical phenomenon, such as the modern science or the scientific revolution that sets its background.

Can we say that the new epistemological condition that emerged by the evolution of scientific developments, in a similar way, necessarily transformed the natural processes into the technical outcomes?[3] Current explanations for the scientific revolution imply that a necessary connection between post-modern "knowledge" and "practice" was established for the first time in the sixteenth century. What we call as "pure science"[4] should be at the same time "applied science" in order to establish a determinative connection between historical epistemological process and its practical space. What is the limits of being "pure" or "applied" for this new epistemological condition? If the pure science is neuter and free from pragmatism, does it have a neuter practice of this sort and, related to that, is a neuter technology possible? If the science is a theoretical *tekhne* (art, technics), then is *tekhne* a practical science? Does every new knowledge obtained from nature and universe, by translating itself somehow to technical processes, transform the human being and his adventure in this world? Does this transformation similarly transform the human being too? To answer these and similar other questions, the relationship between modern science and its connection to the practice offers a significant starting point. Basically, extreme theoretical and practical discussions and parallel assumptions created by the way human beings have relation with nature set the limits of this connection. One can define this dynamic process as considering the human being as part of nature by excluding the mind, accepting the products of the mind, and reflectively making it effective on the mind. Every artificial nature created in the external world, however, affects reflective human beings and therefore the mind. From this perspective, the idea that, just like the scientific revolution, modern science is unique is based on the claim that present condition emerged through a natural outcome of the scientific revolution. Therefore, the question of what the scientific revolution actually was and its answer have the potential to have a direct effect on most of the established assumptions today.

Modern science differs from its origins of "the philosophy of nature" because it is an organization conducted through specialized elements with social functions. Modern societies have complex life experiences guided by "scientists" who have capacity to produce authoritative knowledge by new organizational methods and this philosophy of life is continuously supported and maintained by both the political authority and other elements controlling the market. This kind of knowledge is produced through engineering designs, also called "applied science," that sustain social structure and enable human beings to encounter with nature and it represents the "science's" form beyond the philosophy of nature. Was the source of success we ascribe to the Western science today an apprehension that appeared with the scientific revolution or was it a rational essence that nobody had discovered? What kind of social class does the scientist represent and under what conditions do they produce knowledge?

The cooperation between modern science and daily practical life could, however, be based on the principle of a "continuous dependence" distance from past relations. The relation between the nation-state and citizens is one of the important spheres in which this kind of reciprocal dependence was maintained. Even if this relation is maintained by the necessity of "specialization," it is clearly essential to have for this relation scientists and fields of applied science. The most important parameter for modern life is that scientists are the sole authority that can produce authoritative knowledge on the nature of (artificial or natural) things. It seems that modern science is made of a bowl including the "philosophy of nature" and "usefulness" in its widest sense. As, for example, to the arguments on the philosophy of nature, that the answers put forth to grand cosmological questions, such as the end of universe and the future of this world, seemed satisfactory has been ascribed to its usefulness and success in practical spheres. This fact should be related to the "ontological security,"[5] which can be defined as each human being's positioning and giving meaning to himself/herself within the universe and infinity through an environmental and cultural belonging and mentality in a certain spatial-temporality. When this positioning is consistent, no problem of security appears; the sense of ontological security justifies the individual's relations to all hierarchical elements through the "order of knowledge." The relationship between the modern state and citizens depends on the confidence created through a kind of scientific ideology. The scientific language as one of the most important sources of legitimacy endorsing the bourgeoisie that supported the authority gained power by a creedal level.[6]

The rise of modern science, or its ability to rise, on its own as divorced from philosophy was allowed by the possibility of empirical science and the processes of reducing the philosophy of nature to mathematical formulas and largely by the evolution of the practical to be the determinative element. The transformations within the complex of *magia naturalis, philosophia naturalis and scientia naturalis* occurred due to the practical elements and demands rather than cosmological and philosophical disagreements. The fact that the shared element of the three concepts is *naturalis* shows that the determinative thing in all these epistemological processes is the knowledge on nature related to the ontological security.

In addition, for they are platforms in which these transformations took place, "spaces" and "opportunities" where the epistemology intersects with practical elements are included in the framework of the work. Therefore, starting from the appearance of first effects of the thirteenth-century translation movements until the assumption of universities to play a central role in the relation between theory and practice by adopting the new science in the core curriculum, places like *laboratory* and *Kuntskammer*, epistemological approaches like *magia naturalis*, descriptive background like *imitation* as well as founding individuals such as Galileo, Hooke, Francis Bacon and *epistheme-konomik*[7] (episthemeconomic) structures like "privileged knowledge system"[8] constitute the main framework of this work. A work on the theory-practice relations of the scientific revolution should, of course, be based on this kind of concepts and tools.

To outline theoretically modern science's relation with the practice has a particular importance because it would also show what is unique in the Western world. Answers to the questions posed above will contribute particularly to the theories that explain why others failed while the West succeeded. In addition, this investigation includes significant answers in an inverted way to such questions why nature could/was not expressed in mathematical natural laws or why existing explanatory activities were not maintained in non-Western worlds, such as Chinese and Islamic civilizations.[9]

Besides, according to the classical thought, the human being does not have a space in nature because of their nature, but could have a relation with nature only through knowledge. In modernity, the human being, thanks to their being the active participant in this relation, obtained the right and authority to clarify ultimately all kinds of knowledge by specifying the rules of the game in time—and by living according to physics, biology and economics. Then all social and physical sciences began to be organized in respect to these themes. As for the classical thought, to have an existence that would dare by its nature to know the whole nature within the limits of the world is impossible.[10]

0.1 Conceptual Framework and Tools of Analysis

All theses and assumptions explaining or modeling the formative phases of modern science or the scientific revolution concur on that it emerged out of a kind of break. There is also a concurrence that this break was unique in the history of humanity; generally this fact is meaningful only within the Western experience and its basic dynamics constitute the backbone of the Western paradigm.

Looking at the experimental background of modern science, one can see some transitional spheres. These transitional spheres are, in some sense, the interference areas between theoretical transformations and the practical zone. These are quantification, management and property of privileged knowledge, *magia naturalis*, laboratory as workshop, mechanical arts and philosophical apparatus. Quantification is, in some sense, the beginning of the path for the philosophy of nature to mathematical principles, although under which conditions it is formed and what the quantified quality means as a language has distinct importance. Two phenomena are considered to be critical developments: (i) the question of "quantification of the qualities," which had been considered a taboo until the scientific revolution, became a necessary condition for beginning the revolutionary processes and (ii) the mathematical language gained vital importance in order to have public knowledge and "ontological security." The question of what kind of scientific perspective the quantification foresaw is critical for showing whether it was unique to the Western science or not.

As I consider the laboratory, the place for the work of quantification, as the workshop of modern science, I will look at its distinguishing factors and empirical methods from the tradition of alchemy only in its relation to itself. I will investigate how mechanical arts evolved to a key position in the literature throughout the Middle Ages and how these arts responded to what kind of needs at the end of this process.

The order of privileged knowledge is perhaps the most important parameter of the modern science paradigm; it was involved from very early on in the process of institutionalization and classification of knowledge and kept its effective position. I will look at the encounter between the idea of property rights—in particular intellectual property rights—and the idea of secret thought as well as the designation of a whole paradigm by the patent system.

All these concepts represent the practical processes that require attention in order to answer why modern science adopted practice at the core and why technoscience had a decisive position. While examining these processes, how *artes serviles*[11] turned into *artes mechanica* and how *artes mechanica* turned into a kind of *philosophical apparatus* will be critical to understand the "transformation of *tekhne.*"

Factual analysis tools that I use in this study adopt, in some sense, the "ethnomethodological" method. *Ethnomethodology*, developed by Garfinkel, is a view that identifies human behavior with practical reasons rather than logical models. According to this, the determinants in social facts are the methodological functions and activities that give meanings to individual daily lives rather than the facts that occur beyond the individual or social capacity. Human beings can create facts and realities only through behavioral forms. The practices and experiences within social structures are results of completely pragmatic and rational interests, which should not require certain definitions or any theoretical framework.[12]

Practical conditions form the perception and design of reality. The scientific revolution was too basically built on such a change and transformation and it was not a simply a theoretical expansion. The most important factor questioning the basic system was the "ontological restitution" that appeared out of tension between exogenous practical elements and internal ones. In fact, the process that we call the "scientific revolution" can be characterized as a "ontological construction" based on a new perception of reality caused by exogenous elements. Thus, the Enlightenment philosophers' view of rational and experimental science as special and as a criterion was due to its success to create a new social order.

Although some of the questions I posed from the beginning are not necessarily related to this work, I listed them here because to provide correct answers to them depends on having a proper explanation for the relations between modern experimental science and practice.

Even if the subject of this work is based on philosophical concepts, especially the relationship between the proposed framework and practice requires to consult the sources of the history of science and arts as well as those of the history of philosophy. The sources are selected from the works of those who worked in the workshop of practical arts. The history of technology is not taken as a starting point, because it is not directly related to the questions posed above. The work's focus on the ontological origins of the *tekhne*, therefore, drops the ethical dimensions from the general framework.

Notes

1 Immanuel Kant, *Logic*, John Richardson (trans.), London: W. Simpkin & Marshall, 1819, p. 30.
2 Alexandre Koyré used the concept of *scientific revolution* for the first time in the twentieth century. Even if a historical emphasis on "a scientific revolution" for earlier period did exist, its usage in the modern sense belongs to Koyré. For popular use of the concept see Herbert Butterfield, *The Origins of Modern Science*, Great Britain: G. Bell & Sons Ltd., 1957, p. 7.
3 *Modern science* is a special kind of knowledge that transformed some elements in the whole historical process of theoretical knowledge and reached a teleological stage by eliminating some of them. It is observed that this structure transformed the classical body on the philosophy of nature within the notion of "science" with a new form after the industrial revolution.
4 According to John Horgan, *pure science* is the science produced only for its use. See Brian Czech, "Incorporating Nonhuman Knowledge into the Philosophy of Science", *Wildlife Society Bulletin*, Vol. 29, No. 2, Summer 2001, p. 665–74.
5 *Ontological security*, a concept developed by Giddens, means a sense of confidence in the external world and close circle obtained through individual's practical life. See Anthony Giddens, *Modernity and Self-identity: Self and Society in the Late Modern Age*, Stanford: Stanford University Press, 1991, p. 35.
6 İhsan Fazlıoğlu, "Modern Bilim: Sahih Bir İtikad Arayışı", *Anlayış*, 19 December 2004, p. 18.
7 *Episthemeconomic* means knowledge or its related object that has become or has the tendency to become a commercial commodity and is available to relocation.
8 The rise of intellectual accumulation and artistic skills with the question of property and the resolution of the both simultaneously in a systematic way in this period were the important composition of the new knowledge systematics. Occupational secret knowledge was critical for the survival of

that city-state and was one of the essential conditions for comparative advantage in regional competitions. Generally, just as many Medieval units, these kinds of licences remained within the family but thus did not carry any element encouraging novelties. However, this kind of understanding of intellectual property played a critical role for the emergence of the patent system. First examples of the patent system among Italian states seem to have emerged in Florence. See Pamela Long, "Invention, Authorship, 'Intellectual Property,' and the Origin of Patents: Notes toward a Conceptual History", *Technology and Culture*, Vol. 32, No. 4, 1991, p. 846–84.

9   Many historians of philosophy, such as Zev Bechler, question why the scientific revolution occurred in the West, but not in China or in the Muslim world. These historians argue that generally Islamic theology (*kalam*) can perceive nature quantitatively because it is a mathematical ontology and what the thinkers like Ibn Haytham did was not quite different.

10 M. Foucault, *The Order of Things*, New York: Routledge Publications, 2005, p. 338.

11 *Tekhne* is generally defined also as *artes serviles*, since it fell into the field of work for unfree people in the Aristotelian sense, *epistheme* was always a more noble field.

12 Alain Coulon, *Etnometodoloji*, Ümit Tatlıcan (trans.), İstanbul: Küre Yayınları, 2010, p. 27.

1 Imitation

Whenever an investigation on nature was the case in the history of thought, consulting to the ancient wisdom is indispensable. The Greek legacy, largely referred now as the only literary source, provides the first examples on how the philosophical context on the nature of nature and the nature of human being was established. Besides, important connections on the relationship between *tekhne* (art, technics) and *physis* (of nature) were established as well in this period. Some questions were posed: "How did the idea of this relationship emerge?" "Is this relationship necessary or do *tekhne* and *physis* act independently?" or "Is there any question of limits between *tekhne* and *physis*?" In general, the Greek legacy is considered the starting point for all these discussion(s). The relation or distinction that we try to build between theory and practice is related to under which conditions they were taken and what kind of reflections they had in the past.

1.1 Ancient Relations

Epistheme comes from the Greek root *epistasthai*. *Epistasthai*, to know how to do something, is important to show *epistheme*'s connection to the practice. The application of this word whose exact meaning is *know-how* is best seen in the drawings of military arts. *Technai* comes to learn military arts from the knowing person (*epistamenon*).[1] Socrates asserts that the military arts are innate or they are transmitted through knowledge (*epistheme*).[2] Plato, as the most important transmitter of the thoughts of Socrates, writes about *tekhne* and its relations in many of his books. In Xenophon's Socratic works, *Memorabilia* and *Oeconomicus*, Socrates calls certain arts such as mathematics and astronomy many times as *technai*, and occasionally he calls mathematics and astronomy as *episthemai*.[3]

DOI: 10.4324/9781003275756-2

Xenophon in various parts of his book *Memorabilia* tells the art of government (*basilica tekhne*) without differentiating theory from practice, and he uses *banausikai tekhnai* (vulgar-rude) for *illiberal arts* in *Oeconomicus*.[4]

Plato's works include many details on various arts. Occasionally, Plato identifies theoretical aspects of some practical arts (*tekhne*) such as navigation, military, sculpture, administration and medicine with *epistheme*. In additio(n), he relates *tekhne* with *gnosis*. He divides all sciences into two: *praktikos* engaging in practical things and *gnostikos* as the completely intellectual activities.[5] Both nature and *tekhne* have lower position in respect to the ideas. Both of them seek to realize the ideas, but nature is considered relatively superior than *tekhne*. In this sense, *tekhne* is classified as "real and artificial."[6]

There are four kinds of *tekhne* in the dialogues of Socrates. According to the dialogues of Polus and Calliclas, these are medicine,[7] physical education, legislation and jurisdiction. Other arts somehow fall into these categories and each art is described by its objective. In the texts, *tekhne* and *epistheme* are mostly used interchangeably.[8]

Even though Aristotle classified *tekhne* among practical science, he emphasized its character of deriving practical knowledge and rejected the Platonic approach.[9] In his book, *Nicomachean Ethics*, where he discussed classifications, Aristotle divides mental virtues into two: *logistikon* (calculation) and *epistemonikon* (science). At large, *logistikon* deals with daily affairs and changing facts, while *epistemonikon* corresponds to the unchanging structures like mathematics. As calculation deals with the practical sphere, differentiating "good" and "correct" from "right desire" appears in this category. Similarly, in the differentiation between *tekhne* and *virtue*, he states that *tekhne*'s aim is for solid product whereas *virtue*'s objective is for activities, such as playing an instrument.[10]

In addition, according to Aristotle, thought has five virtues: *Tekhne, Epistheme, Phronesis, Sophia* and *Nous*.[11] At the point where he distinguished *epistheme* from *tekhne*, he states that *epistheme* deals with constant things. Therefore, *epistheme* is completely outside of *tekhne* and deals with the unchanging facts. In this context, he indicates in *Metaphysics II* that the matter in the external world and changing things will not be subject matter of *epistheme*; thus, we should expect less in this field than in unchanging information.[12] He also specifies that *epistheme* may cover the facts that happen regularly in the nature and the things that are mostly realized.[13] Appealing to *tekhne* to understand nature, Aristotle concludes that nature is teleologically indigenous:

Where there is an end, all the preceding steps are for the sake of that. [...] Now action is for the sake of an end; therefore the nature of things also is so. Thus if a house, e.g., had been a thing made by nature, it would have been made in the same way as it is now by art; and if things made by nature were made not only by nature but also by art, they would come to be in the same way as by nature. The one, then, is for the sake of the other; and generally art in some cases completes what nature cannot bring to a finish, and in others imitates nature. If, therefore, artificial products are for the sake of an end, so clearly also are natural products. The relation of the later to the earlier items is the same in both.[14]

Aristotle here states that just like artificial objects, nature too continuously proceeds to a certain goal. In both *tekhne* and *physis*, every stop is a starting for the next one. Based on the discussion above, however, it can also be argued that *tekhne* is more advanced than nature. As nature do not make it, *tekhne* made house and boat; and if nature did so, it would do exactly the same in the same fashion. It can also be said that *tekhne* eliminates nature's deficiencies. Contrary to what may be supposed, there is a complementary relationship rather than a contrasting one. In addition, he comments on *tekhne*,

All Art deals with bringing something into existence; and to pursue an art means to study how to bring into existence a thing which may either exist or not and the efficient cause of which lies in the maker and not in the thing made.[15]

The reduction of the distinction between nature and art to whether the source of change is indigenous or exogenous leads to think natural and artificial processes in parallel terms. The subject is discussed in *Metaphysics* at the paragraph z7–9. In both generations (nature and *tekhne*), matter has a form. The form in the artificial one is in the artist's mind, whereas the one in nature is in parents. In fact, even though Aristotle seems to separate natural and artificial environments, and ultimately parallel outcomes by products emerge.

According to a work, *Mechanical Problems*, attributed to Aristotle, there are two extraordinary things: the things occurring by nature with no known cause or origin (*kataphysin*) and the things occurring by *tekhne* (*paraphysin*). The human being whose strength, knowledge and capacity is limited grows old due to nature's hard and coercive structure. Intervening at this state, *tekhne* transforms the loss into benefit; the part of *tekhne* that facilitates this is called *mechane*.[16]

Moreover, the facts happening indigenously fit to nature, whereas the exogenous facts happening by force are contrary to nature; in other words, they are *paraphysin*.[17] Therefore, the characteristics of being contrary to nature and the one of occurring by force are used interchangeably. However, this should not mean a break from natural processes. For example, lightning occurring from heavens contrary to the nature of fire is *paraphysin*.[18] Aristotle uses in some of his works the concept of *paraphysin* for the great birds rarely seen in nature too.[19]

Even though nature and *tekhne* seem to be opposed to each other, this conflict actually appears in the human being. The human being's sophisticated needs clash with the simple condition of nature, then *tekhne* saves the human being from despair by resolving the conflict. Therefore, the adaptation of the animate is facilitated similarly by *tekhne*.

Mechane meant in the fourth century BC to overcome the difficulties in a crafty or by tricky way by using mental or physical tools.[20] It can be said that these tools are as a whole beyond the divisions of natural or artificial categories. Like Prometheus, it is said that the human being struggles desperately in nature but then overcomes the challenges by the discovery of *technai*.[21] Nevertheless, despite the discoveries gathered from all the mortals, the human being cannot escape from the control of God Zeus and "complains that he has no *sophisma* for extricating himself from the present predicament in spite of all the inventions (*mechanemata*) which he devised for the mortals."[22]

The author continues in *Mechanical Problems* to explain the characteristics of *mechane*. Mechanics is a general way by which heaven burdens are lifted by small forces and the small manages the big. In this context, while the management of the small forces by the big forces is natural, by *mechane* the small forces come to manage the big forces. In other words, the relationship between forces is inverted. Citing Antiphon, he states that "We by skill gain mastery over things in which we are conquered by nature."[23]

Aristotle next discusses that nature works like an artist (*mechane-mechanastai*). The movements of the animate resemble to the movements of puppets that are some sort of mechanical tools.[24] Besides, epistemologically he refers again to the mechanics while listing certain sciences such as optics and astronomy. He states that just as the application of arithmetic proofs to harmony, so too geometric proofs can be applied to optics and mechanics,[25] and *tamechanica* is a subdivision of stereometry (geometry's subdivision measuring solid bodies).[26] It is true that Aristotle is not so strict in delimiting *epistheme* and *tekhne* although he stays loyal to Platonic differentiations.

For example, he defines medicine as *epistheme*, and occasionally he uses both concepts together. In addition, *tekhne* is not something that can be qualified by theoretical and practical knowledge in Aristotle's classification.[27]

The differentiations on *tekhne* and *epistheme* appears utterly in the Stoic period. Zeno mentions a kind of *tekhne* to cure the diseases of the soul.[28] According to him, *tekhne* is a systematic collection of cognitions. Chrysippus, the one considered to be the second founder of Stoicism, states that practical judgment (*phronêsis*) is a kind of *tekhne* concerning the things having to do with life.[29] According to Stobaeus, the difference between *tekhne* and *episteme*, properly speaking, is that the latter is said to be secure and unshakeable by reason. *Tekhne* does not have the same kind of stability.[30]

Although it has had many transformations in Ancient Greek thought and textual source, we can summarize the general conception of *tekhne* in the following way: (1) Even though there is no complete agreement on its definition, *tekhne* is actually the collection of measures, including everything discussed above, that starts with the human being, endures in nature and balances nature. (2) *Tekhne* can be a thing itself, the knowledge of that thing, the subject of that thing. As it can be understood from the examples given above, *tekhne* has a wide field of usage from handicrafts to intellectual arts.[31]

1.2 *Tekhne* as Imitation of Nature

Tekhne, being in some sense human beings doing things, is also an ordering or an intervention on, the "existent" things. This ordering and intervention, however, is a human "contribution" rather than an enforcement to nature; it shares the same space as part of the member of the same world of existents and enters into reciprocal "relationship" within this shared world. From this point onward, the possibility and way of human "doing" becomes important. Therefore, the possibility and way of human "doing"—*tekhne*—is also important for digging out the historical and philosophical background of technology.

Debates on human limits from the Greek philosophy evolved, in Roman thought, also fostered by Hermeticism, into the discussion of "the possibility of human divinity." This idea first appeared, for example, in Plato's philosophy, in a context of resembling God to an architect and of discussing whether human being can be God. Similarly, in Greek theater, the fact that *mechane* takes to the stage by presenting God as *Deus ex machina* shows that God also acts as a mechanist.[32] In addition, Cicero referred to the divine character of human "doing" by saying that Archimedes attributed to spheres the universe and planets that Plato's God, in *Timaios*, made and he could adjust these spheres' motions without a divine soul.[33] As for Aristotle, all human doings (*tekhne*) are a kind of existing, participating in "existence."[34]

In his famous comparison, Plato explains the origins of life and material world as,

> So long as its wings are undamaged, the soul travels through the heavens; but some souls lose their wings, fall to earth and take to themselves earthly bodies. There follows a vivid picture of the procession of souls, headed by Zeus, to the rim of heaven, and of the difficulty experienced by the human souls in following the divine.[35]

A similar explanation is true for *tekhne* too. Like Plato's God Demiurge, Archytas vivified the matter: "... for many men of eminence among the Greeks [...] that the model of a pigeon formed in wood by Archytas, was so contrived, as by a certain mechanical art and power to fly ..."[36]

According to most of early philosophers of the history of thought, the examination of nature is actually the examination of matter.[37] As for Aristotle, there are two things called as nature in the Ancient Greek thought: form and matter. Therefore, if *tekhne* imitates nature, it is part of the same discipline to know form and matter.[38]

> In all other things which involve production for an end; the product cannot come to be without things which have a necessary nature, but it is not due to these (except as its material); it comes to be for an end.

As nature has two layers, nature as matter and nature as form, based on the principle that What is necessary then, is necessary on a hypothesis; it is not a result necessarily determined by antecedents. Therefore, "necessity is in the matter, while 'that for the sake of which' is in the definition."[39] Similarly, concerning the characteristics of *tekhne*, he says "the principles by which things are made are inside those things, they come from the minds of artists"[40] and "tekhne carries the capacity to make in its inside, its existence and origin is its maker."[41]

Mechanics is considered by many philosophers as one of the superior occupations. This science examines not only the objects moving by their nature but also those moving by force—in other words, contrary to their nature. Therefore, all systems applying force with a certain objective or

functioning through forces are among subjects of this science. The followers of the school of Heron classify the mechanics' theoretical side
as geometry, arithmetic, astronomy and physics and its practical side
as carpentry, blacksmithing, construction and dyeing. Among all these
arts, the most superior ones are those called "mysterious workers" who
make military arms and machines working by force. Besides, the mechanical systems that were most advanced in their times are observed
in Archimedes's *On Floating Bodies* and Heron's *Hydra*. These systems
are mostly mechanisms in which heavens are represented in kinematic-
geometric facilitated by hydraulic movements and energy.[42]

According to ancient understanding of mechanics, the human being
makes machines by imitating nature in matter, form and even motion
observing the external world. The thing that gives motion to these machines shows that the human being has a mental ability. As it can be
seen here, it is the power of thought that designs all these tools although it may seem invisible at first. The human being exhibits his
nature's mental power by using the matter in that way. Thus, as *tekhne*
is a consistent thought leading to an end by matter, the thought (*logos*)
is the thing giving motion to the mind.

Gregory of Nyssa's following statement is critical for showing the
limits of human mind:

> The skill applied to the mechanisms all but serves instead of a soul
> for the material object (through these mechanisms the object im-
> itates motion and sound, various shapes and the like) would be
> proof that there is something in man of the sort that through imag-
> inative and inventive power is able to understand machines within
> itself and to plan them in thought, and then to put them into action
> by skill and to manifest the thought in matter.[43]

The notes collected from the history of philosophy-science shows that
a kind of scaling between God's "knowledge" and "creation" on the
one hand and human "knowledge" and "making" on the other is applied. Eventually a kind of modeling based on the *imitation* of nature
can be foreseen.[44] A line coming from Plato to Stoic philosophers
adopted matching and even identification God and artists, nature and
art (*tekhne*). This modeling is used both for macro cosmos, the universe, and for micro cosmos, the human being.

In fact, to explain Aristotle's view on the movement of the animate
in *De Anima*, Jean Buridan takes it a little further and makes an analogy between "clock" and the "movement" of the animate. He com-
pares clock's hairspring and strap with human muscles and tissues,

and finds parallelism between the movements of the two by likening human bones to the clock's wooden or iron materials.[45] He states that God gave an *impetus* to the heavenly bodies, and just as a mill's propeller cannot be stopped so easily because of the *impetus* it received form winds, so too the *impetus* given to the heavenly bodies enables them to have motion eternally.[46]

As these short examples show, many philosophers find various connections between God's creation and human making. Starting from the sixteenth century onwards, "nature" began to be seen as an "instrument" (*natura machinata*) and it was accepted that the human being can make, once he wills, his own *machinatio*, because just like God creating "nature" human being has a divine intellect too.[47] Therefore, a distinction is made between human arts and the "machine" as collection of mechanisms; the emphasis is concentrated more on the machine. It is underscored that "nature" discloses various examples of animate and inanimate mechanisms, and the human being can make similar machines, in other words, mechanisms, by looking at these examples disclosed and at the view of heavenly bodies as one of the premises of the ancient cosmology. These "makings" based on imitation is clearly a novelty for the human being.

In addition to being an outcome of a four-century-long process, the seventeenth-century modern science paradigm gained a characteristic nature by having developed in the conditions in which ancient imitative ontology motivated. Ancient legacy showed itself as a ground of description for social dynamics and struggles of class and power as well as an intellectual background. Therefore, it is possible to say that "imitation" was one of the most important parameters shaping the modern science circles fundamentally.

Notes

1 Xenophon, *Memorabilia*, E.C. Marchant (trans.), Cambridge: Harvard University Press, 1923, p. 147, 171–3, 283.
2 Xenophon, *Memorabilia*, p. 171.
3 Xenophon, *Memorabilia*, p. 171–2.
4 Xenophon, *Oeconomicus*, E.C. Marchant (trans.), Cambridge: Harvard University Press, 1923, p. 391.
5 Harvey D. Scodel, *Diaeresis and Myth in Plato's Statesman*, Göttingen: Vandenhoeck & Ruprecht, 1987, p. 35.
6 Plato, *Republic*, Tom Griffith (trans.), Cambridge: Cambridge University Press, 2000, p. 18–20.
7 In many Hippocratic texts, medicine is described as a *tekhne* gathering information on the human nature. It is also held that the thing heals the human being and turn human to normal condition through healing is the

human nature, and the physician only helps this process. In this context, *tekhne* is an inquiry about the human nature. For Hippocratic *tekhne* see Heinrich von Staden, "Celsus as Historian", in *Ancient Histories of Medicine*, Philip van Eijk (ed.), Leiden: Brill, 1999, p. 251–95.

8 Plato, *Gorgias*, Benjamin Jowett (trans.), USA: Agora Publications, 1994, p. 7–24.

9 Aristotle, *Nicomachean Ethics*, 2003, p. 333.

10 Aristotle, *Nicomachean Ethics*, 2003, p. 327.

11 Aristotle, *Nicomachean Ethics*, 2003, p. 331–2. *Phronesis* is a concept in Classical Greek thought used by Aristotle like Plato in the sense of "Practical wisdom" in order to distinguish it from theoretical wisdom (*sophia*). In its widest meaning, it is practical wisdom put forward for the capacity to use right tools to obtain the best results for oneself or for a happy life.

12 Aristotle, *Metaphysics*, John McMahon (trans.), New York: Prometheus Books, 1991, p. 46.

13 Aristotle, *Metaphysics*, p. 127.

14 Jonathan Barnes (ed.), *Complete Works of Aristotle: Physica*, Princeton: Princeton University Press, 1995, 199a8–20.

15 Aristotle, *Nicomachean Ethics*, 2003, p. 335.

16 Aristotle, *Mechanical Problems*, Cambridge-London: Loeb Classical Library, 1936, p. 331.

17 Aristotle, *On the Heavens*, Stuart Leggatt (trans.), Warminster: Aris & Phillips, 1995, I,II,300a23.

18 Aristotle, *Meteorology*, E.W. Webster (trans.), Kessinger Publishing, e-book, 2004, 342a12–16.

19 Aristotle, *Aristotle's De Partibus Animalium I and, De Generatione Animalium I*, David M. Balme (trans.), Oxford: Clarendon Press, 1972, p. 16–21.

20 In daily language it means "lever" but generally it is the name of mechanical tool rendering movement in a theater stage.

21 David Wolfsdorf, *Plato and the Crafting of Philosophy*, Oxford: Oxford University Press, 2008, p. 105. Prometheus taught fire as the source of all *technai*.

22 Eric Voegelin, *Order and History: The World of the Polis*, USA: Louisiana State University Press, 1956, p. 258.

23 Aristotle, *Mechanical Problems*, p. 331.

24 Martha C. Nussbaum, *Aristotle's De Motu Animalium*, Princeton: Princeton University Press, 1985, p. 345.

25 Aristotle, *Posterior Analytics, The Works of Aristotle*, G.R.G. Mure (trans.), USA: Oxford University Press, 1928, 76a23–5, Ch. 13.

26 Aristotle, *Posterior Analytics*, p. 78b36–9.

27 Aristotle, *Metaphysics*, 982Ia,982bI0; Aristotle, *Nicomachean Ethics*, 2003, p. 333–5.

28 Richard Parry, "*Episteme* and *Techne*", *The Stanford Encyclopedia of Philosophy (Fall 2008 Edition)*, Edward N. Zalta (ed.), http://plato.stanford.edu/archives/fall2008/entries/episteme-techne (10 November 2010).

29 Parry, http://plato.stanford.edu/archives/fall2008/entries/episteme-techne. *Phronesis* is also a kind of *tekhne*, and foresees to make right and good things concerning life. Aristotle's distinction has more general meaning other than judgment.

30 Cicero, *On Moral Ends*, Julia Annas (trans.), Cambridge: Cambridge University Press, 2001, p. 80–1.

31 In German *die hohe Künste* means "high arts" and used for philosophical or intellectual activities. Heidegger identified *tekhne* with this concept for its connection to *epistheme*.

32 It is the Latin version of Greek expression Θεὸς ἀπὸ μηχανῆς—*Theos apo mēchanēs* and tells that an impossible final stage of a theater could only be possible by the God's mechanical intervention. It is important for showing the ancient origins of the divinity in modern imitation. However, Aristotle criticized the insertion of *mechane* into the play with such an intervention and argues that all applications should come as a continuation of the previous scene. See. Aristotle, *Poetics*, New York: Penguin Books, 1996.

33 Cicero, *Tusculan Disputations*, C. D. Yonge (trans.), 1877, Ebook: New York: Eco- Library, 2007, p. 36.

34 The concept of *mimesis* appears in every phase of the classical Greek thought. However in Platonic or Aristotelian philosophies it has different meaning. In fact the cosmologies in both systems were different. For Plato, the imitation has no importance, because the original one was the idea, that is, the thing imitated. Aristotle, however, argued that the human being would increase their knowledge by this imitation.

35 Plato, *Phaedrus*, R. Hackforth (trans.), Cambridge: Cambridge University Press, 1952, p. 69.

36 Aulus Gellius, *Attic Nights*, Vol. II, W. Beloe (trans.), London: St. Paul's Churchyard, 1795, p. 223.

37 Early philosophers were called as the philosophers of nature. They accepted fire, water and air as the origins of existence.

38 Aristotle, *Physics*, South Dakota: NuVision Publications, 2007, p. 28.

39 Aristotle, *Physics*, p. 38.

40 Aristotle, *Physics*, p. 131.

41 Aristotle, *Nicomachean Ethics*, 2007, p. 95.

42 John W. Humphrey, John P. Oleson and Andrew N. Sherwood, *Greek and Roman Technology*, London-New York: Routledge, 2009, p. 46.

43 Gregory Nyssa, *On the Soul and Resurrection*, Cathrine P. Roth (trans.), Canada: St Vladimir Seminary Press, 1993, p. 42.

44 *Cosmo Mimesis*: conducting activities by taking the universe as reference point and making inference from the ideal conditions there. Although *mimesis* evolved into various meanings in Ancient Greece, generally it widely meant representative imitation of nature. Plato used three bed metaphor in a dialogue with Socrates in his *Republic, Book X*: the first is in the God's mind (*idea*); the second was the bed produced by the carpenter, which was an imitation of the one in the God's mind (*mimesis*); and the third was the artist's imitation by drawing of the carpenter's table (*mimesis*). See Plato, *Republic*, Tom Griffith (trans.), p. 315–6.

45 A.C. Crombie, *Styles of Scientific Thinking in European Tradition*, Vol. II, London: Duckworth, 1994, p. 1091.

46 Jürgen Sarnowsky, "Concepts of Impetus and History of Mechanics", in *Mechanics and Natural Philosophy Before the Scientific Revolution*, Walter Roy Laird and Sophie Roux (eds.), Dordrecht: Springer, 2008, p. 134.

47 Vitruvius, *Ten Books on Architecture*, Morris H. Morgan (trans.), Cambridge: Harvard University Press, 1914, p. 328.

2 The Epistemological Transformation of Secret Knowledge

From the thirteenth century onward, translations from Arabic began to be influential in the whole continent. Until then, Neo-Platonism and Hermeticism—which were not represented in the university curricula—paved the way for pervading of the belief of "an unavailable and different knowledge, which was also endorsed by the translated texts. Especially the Christian conquest of Toledo in the late eleventh century allowed the Muslim Spanish legacy to be transferred to the Western world. The knowledge in Muslim Spain had aimed at establishing social order in natural ways and institutional enterprises welcomed philosophy and natural sciences. Since large-scale syntheses of transitions based on Aristotle's or Plato's philosophies were made, it was possible to have an "order of knowledge" based on knowledge-society relations in which explicit definitions on natural sciences and the social reflections of these definitions were highly prevalent.

Thanks to the fact that the Muslim societies translated the new forms of theory-practice relations into the world of realities, mathematical sciences paved the way for commerce and accounting, so did geography and astronomy for the freedom of travel, so did alchemy for a rich material culture and so did botanic and zoology for highly detailed agriculture and husbandry. This rational relationship between theory and practice fed each other,[1] in a process of "freeosmotic."[2]

Since the private life or the idea of privacy did not exist in the Western geography until the mid-sixteenth century, the hierarchical order precluded the emergence of the individual or allowed only an idiosyncratic meek individualism to appear. Living spaces were limited places generally encompassing mid-range relatives and having passages from one to another instead of independent rooms.[3] By virtue of the Protestantism for the first time, a house with independent bedrooms connected to each other by a corridor would show up. How would the

DOI: 10.4324/9781003275756-3

individual emerging under these conditions express her/himself first in the family and then in the outer society?

The solution that would enable the individual both to exist in the public sphere and to be self-sufficient foresaw an epistemological order consistent to overcome the arising crisis. This order permeating into the all social processes was being protected by a legal shield and the individual was transforming into the smallest unit of the homogenous society, which had hitherto been unseen. The facilities of spreading knowledge increased by the power of individual expression that was strengthened in an ontologically secure ground, thereby knowledge was able to appear without falling into a disruption based on assumptions dismissing class structure. Thus, the society could transform the potential of knowledge increasing in a liberal order bound by local and regional economic trends into a form of life and representation, and this form could evolve into an important accumulation of knowledge that could be applied to wide range of fields from medicine to astronomy. However, the transformation of this accumulation into institutional structure internalized in the Western world took place in such a long period of 300 years. The "idea of secrecy" that can be defined as a transcendent world of reality of which the individual in the society never thought to have grasp had different meanings for the Western and the Muslim societies.

As the texts as source of mysteries(secrets) penetrated into the Latin world, it grew to be more foundational, because the Latin thought on the ancient stood in a meta-historic context. Therefore, an ontological connection between the ancient and the contemporary could not be established on a rational ground and the perception of secrecy largely persisted. The unique source of richness and order in the Muslim world was "knowable world of secrets," which was perceived as a hidden treasury beyond the observable things. Translations in this context had a completely homogenous and hybrid nature made of Hermeticism, Platonism, Greek thought, Mesopotamian thought and Indian thought.[4] Secret claims found quick reception because a hierarchal relation was not established among these texts and sound texts were not distinguished from forged ones. For the Western world, everything unknown was secret, because the right to access on and management of the Scholastic thought and the knowledge offered by the Church was not sufficient to overcome ontological questions.

Undoubtedly, these exciting texts had completely reflections in Islamic science because they were detached from their original contexts—in fact Muslims distinguished Hermeticism from Hellenistic mythology and mysticism. The Islamic philosophy as a whole, however, is based on the idea that nature or the universe is made of secret harmony or of temporarily unknown occult forces and the only way to understand them is by means of knowledge. In addition, while it is apparently clear that the thing that opens up all these obscurities and mysteries could not be anything other than knowledge, for the West this mysterious structure devoid of hierarchy itself precludes knowledge. But the system begins to turn the other way around after the fifteenth century, the persons who master these secrets prioritize the practical outcomes rather than the origins of this knowledge thanks to their privileges in the socioeconomic system.

The work *Secretum Secretorum*, whose title is a translation of Arabic *sirr* secret and *sirr al-asrār* (secret of secrets) was the hitherto most influential and the most famous book in Medieval Age, according to Thorndike, and it had about 600 manuscript copies.[5] Even though it had an original Greek manuscript copy comprising of sections on politics and ethics, its scope expanded to include medicine, numerology, alchemy, witchcraft throughout the Medieval period. Both kinds were translated into the Western world.[6]

One of the most important philosophers who translated and commented on *Secretum Secretorum* in the Western world is Roger Bacon. Bacon relates that his *scientia experimentalis* is the method he used for testing doubts coming from these mysterious structures. By separating legitimate experimental science from witchcraft, he blames the witches for deception and claims that "Art using Nature for an instrument is more powerful than natural virtue, as is to be seen in many things."[7] Bacon's perspective is actually interesting for showing which methodology that the Islamic alchemy books or *sir al-asrār* literature is based on. In the introduction of his work, Bacon states that his work is a gate to "true sciences" by means of "arts helping nature" such as astrology, alchemy and physiognomy.[8]

Being one of the most highly esteemed books throughout the Medieval Age, the work becomes one the sources of motivation breeding modern science with its message of "Knowledge is power." *Secretum* literature claims that one should master in all sciences, especially in arithmetic, which helps understanding nature. Bacon states that the content of the book is written for the experts, and if it falls into the hands of ordinary people, especially the instructions for making gunpowder—Christianity will face the danger of destruction.[9] Acquaintance with secrets of Bacon and other philosophers writing on alchemy is limited to this framework. In the following centuries, however, the concept of *secret* acquired a wide range of circulation and grew to constitute a collection of knowledge sought by all experts and laymen and used for survival.

Throughout the whole Medieval Age, transmission and spread of knowledge was based on two sources: first is the Scholastic enterprise that the universities, respecting the Church authority, transmitted in classical classification; and the other is the collection of secret knowledge and mysteries that disclosed the information outside of the classical classification and carried the characteristics of Hermeticism and Neo-Platonism through certain kinds of texts like alchemy.

The first source, *scientia*—defined as science by the established epistemological system, that is, universities—was based predominantly Aristotelian classifications and definitions. Scientific knowledge was defined as the capacity of revealing the unqualified knowledge.[10] Not only experimental knowledge of facts but also showing their "causes" was prominent. Demonstrative knowledge was defined as showing that the causes of a fact could not be anything else.[11] Scholastic *scientia* was interested in "explaining facts through causes" rather than in experimental results. The simulation of individual results to the universal system was not accepted as scientific product. The second source, the literature of secret sciences, did not conform this classification and it was transferred through completely different dynamics. This literature was based on producing knowledge on the sources of phenomenal facts, also called *occult*, that could not be felt by five senses.

In principle, these two sources grounded in the same way the legitimacy of practical arts. Since the products of these arts were in the literature of secret sciences they were called as part of magic. As *tekhne* (art, technics) was not interested in the objects' existence, necessity and relations to nature, it could not produce scientific knowledge anyhow; thereby, mechanical and technical processes could not yield scientific knowledge either.[12] Thus, *Secreta* could be experienced but could not be explained by the principles of deductive logic or of the classical philosophy of nature, because it was not proper scientific knowledge. Artists, experimentalists, alchemist could produce high caliber knowledge on *facts* whereas they could not produce *causes of facts* and if they did so, this knowledge could not be scientific.[13]

The fact that the literature of secret sciences maintained such a wide range of circulation created great tensions in time. The scholastic authorities preferred to seal the books in order to distinguish the knowledge conforming to *scientia* as product of their own systems. However, the distinction between reality and fallacy in the scholastic order was not based on objective grounds or *scientia experimentalis*; the transmission knowledge was allowed under most special cases for the permanence of the order.

According to the classical classification, the subjects whose education was allowed in universities were *artes liberals* (*trivium* and *quadrivium*). But those who were not free, that is, slaves and lower classes, did not have access to these subjects. They were mostly satisfied with the *artes serviles* (*mechanical arts*—arts for slaves and servants), the occupations that people worked with their body strengths.[14] On the other hand, those who were inferior to the superior scholastic authorities had access, by using unsystematic methods, to the literature of secret knowledge circulating around. In fact, Roger Bacon asserts that even God did not choose to communicate with the noble and superior classes but rather with ordinary people, and he himself learned many things from people rather than books.[15]

In this context, an appropriate example for a person from the Scholastic circles who had works on Medieval arts could be Theophilus. He provided information on handicrafts, glasswork, dyeing and so on that were prevalent in the twelfth-century churches.[16] Unlike contemporary works, the information he gave on these arts was systematic. The framework that Theophilus offered for lowly arts confirmed that they were part of a sacred order. The conclusion we can draw here is that Theophilus who brought a different approach to the literature of secret sciences claimed that this literature should be normalized for the public good.

For whole Europe, one of the factors leading to a new science and order of knowledge almost synchronically in many vernacular languages was the Mediterranean. The Mediterranean, connecting Europe and standing somewhat as a "virtual state," ensured this grow and mature with its warm climate. The thousands of years of legacy of the Fertile Crescent could be transmitted unrestrictedly especially to the European coastal cities with no restriction through small and patient waves of the Mediterranean's calm waters. If Muslim Spain was to be considered the starting point for this interaction, all Southern European coasts became grounds for those seeking solutions to their questions and problems. Thanks to this fact, regional powers obtained an escape point and place of logistic site away from the dominant order. Looking in this perspective at the work of a wine merchant from the fourteenth century, known as Gottfried of Franconia, would bring a surprising richness. The existence of occasional German terms in the text shows that the book paid attention to the vernacular applications. The author tried to suffice the practical needs by collecting information and methods he learned from the oral culture. Owing his experiences early to his travels, Gottfried traveled especially Southern Italy, which carried the eclectic nature of the Mediterranean basin.[17]

Among important books working on theory-practice relations in the Medieval Age was the ones on metallurgy and mining. As these works were largely under the influence of the methodology of alchemy, they offered more efficient results than the results by normal methods. A blacksmith would start such a book with these words: *Nv spricht meister alkaym...*[18] The blacksmith would claim that the conventional method for hardening the iron was to plunge it into cold water but this would not be sufficient for high quality tools and in order to have high quality product one would need to blend water with vegetable oils and animal fats—such as grease for scythe, blood of a billy goat for files and caterpillar for quarry hammer.[19] The metallurgical formulas and blended materials in the recipes were not accidental examples but their selections were related to the argument that each one of their particularities would be transmitted to the characteristics of the produced stuffs. The belief that the blood of a billy goat would also cut diamond shows the cults from the ancient Western texts to Pliny's works.[20]

The works on metallurgy and mining, especially endeavors to find locations and amounts of precious metals were principally important. In his book titled *Pirotechnia*, Vannoccio Biringuccio claims that fortunetellers found mines at mountains by conjuration and asks if this was to be true why they would not use these skills for purification or processing mines. Someone who has capacity for something should have the capacity for similar things too.[21] Biringuccio states that he never denigrates fortunetellers but he himself would prefer a method that was obtained by the experiences of great masters and by the mercy of nature as well as disregard fallacies impossible to happen, and then he would reach a certain result.[22]

In the fifteenth century, especially due to the geographic discoveries, the need for qualified persons carrying certain technical skills burst and more comprehensive texts with specific objectives were sought. Lifting mechanisms and water handling systems entered into the standard list of needs for every entrepreneur, especially in the mining pits dispersed on the whole continent. Existing technical staff in the army—masters of weaponry and tools—were civilized, socialized and began to spread. Quickly changing spaces and high mobility led certain conditions challenging and accelerating the normal economic order such as building castles, making bridges and opening ways and mine up in a short period of time. Discovering more universal methods to transmit their knowledge accurately, the technicians began to use the simplicity of the art of drawing to explain mechanical set ups. Therefore, personal designs that can be transmitted to others and highly qualified skills emerged. The way going to the perspective technique began by this communication pressure.

Many books known as *Kunstbüchlein* containing information of handicrafts for domestic use such as cosmetic dyeing appeared in Germany during 1520s and 1530s.[23] These kinds of works were collections of hand notes gathered together from various workshops by the printers but they were presented to public as the development of arts by scientific methods.[24] However, gaining prestige by writing book could end up with the inventor's economic bankruptcy; as once the book was published all mysteries would become disclosed, and once personal information was publicized, it would lose its power.

Besides, the *Secrata* literature, for the first time after Roger Bacon, began to draw attention around 1555 due to its promise of high reputation. Some private secretaries who claimed to know these mysteries began to produce similar works. The books of some people like Alexis Piedmont succeeded to enter among famous books by having new prints every year.[25] Alexis spent his entire life working on his book, he gathered countless information by traveling to various regions of the world throughout his life. Many travelers wrote similar books containing the mysteries and experiences that they picked from ordinary people and farmers on daily life and health in the Middle East and the Levant. The scope of Piedmont's book extended from tooth shining to perfumery, from jam and preserve to hair dyeing, from daily needs of the middle class to dyeing, alchemy and metallurgy. His work was translated from Latin to almost all European languages.[26] The spread and concentration of daily needs was interesting for showing the level of practical life in Medieval Europe.

The spread of Secreta books in such an extent would find reaction in the society. A priest Tomasso Garzoni (1549–1589), in his book of occupations, mentioned a new group as *professori secreti* among hundreds of occupations. He stated that they were continuously after occult or mysterious things.[27] The most important professor of secrets mentioned in the book was Della Porta. His fame spread to whole Europe thanks to his work *Magia Naturalis*, he wrote in 1558 while he was so young. Even contemporary philosophers of nature like Kepler praised his works (Figure 2.1).[28]

Principal starting point for professors of secrets was their apathy to the relations of the knowledge they transmitted to the theoretic-philosophical systems but their presentation of the results irrespective to the causes and the experiences they heard from others. The prescriptions and recipes in these works spread by translation to many languages at least four centuries. Besides, the tendency to call the

NATURAL

MAGICK

BY

John Baptista Porta,

A NEAPOLITANE:

IN

TWENTY·BOOKS:

1 Of the Causes of Wonderful things.	11 Of Perfuming.
2 Of the Generation of Animals.	12 Of Artificial Fires.
3 Of the Production of new Plants.	13 Of Tempering Steel.
4 Of increasing Houshold-Stuff.	14 Of Cookery.
5 Of changing Metals.	15 Of Fishing, Fowling, Hunting, &c.
6 Of counterfeiting Gold.	16 Of Invisible Writing.
7 Of the Wonders of the Load-stone,	17 Of Strange Glasses.
8 Of strange Cures.	18 Of Statick Experiments.
9 Of Beautifying Women.	19 Of Pneumatick Experiments.
10 Of Distillation.	20 Of the Chaos.

Wherein are set forth

All the RICHES and DELIGHTS

Of the

NATURAL SCIENCES.

LONDON,

Printed for *Thomas Young*, and *Samuel Speed*; and are to be
sold at the three Pigeons, and at the Angel in St.
Paul's Church-yard. 1658.

Figure 2.1 English translation of *Magia Naturalis*/Porta Giambattista, della, *Natural magick*. London: Printed for Thomas Young, and Samuel Speed, 1658, retrieved from https://doi.org/10.5479/sil.82926.390880 02126779.

philosophical ground on which these effective and practical methods were based as "natural magic" continued. In fact, it was believed that all these useful things and practical elements were based on natural and mostly occult forces. Natural magic, *magia naturalis*, was in a position to use all these natural forces and imitating them for the goodness of the humanity.

Modern science's character of being open to the general public has become one of its important features distinguishing from other philosophical systems. According to John Ziman, what makes modern science valid is not from the moral strength or literary skills of the person who produces it but rather a whole scientific circle knows and accepts it. This kind of "agreement" on knowledge depends on only the free accessibility of knowledge by "people" or "public."[29]

In short, while Medieval academy was open to a certain group through scholastic tradition, it was closed to the larger part of the society. The academia turned, in a strict sense of the word, into a noble class. The secret knowledge, which was managed by the scholastic thought and whose access by lower classes was considered to be objectionable, was rendered open to public after the copyrights and privileges of the sixteenth-century authors and inventors were guaranteed.

2.1 Magia Naturalis

The fact that the whole alchemy tradition circled around the concept of *naturalis* is an important and characteristic feature of the Western thought. Even in the sixteenth century, the things sought and lost in the Western world were both natural and magical. This concept formed by Della Porta was taken by most of the historians of science through *naturalis* and it was considered separate from the magic in the classical metaphysics. Della Porta states that the aim of *magia naturalis* was nothing but to examine nature.[30]

Thorndike argues that alchemy and magic provided empirical data to modern science whereas Rossi and Garin see them as precursors to modern science. Even though Rossi and Garin do not have positive view toward *magic* in particular, the context they imagine is *magia naturalis*.[31] In this sense, *magia naturalis* provided important clues to the ontological crisis of Medieval Europe by both its natural and magical sides. *Magia* was actually a metaphysical background sustaining natural and real daily processes. The knowledge on nature could find an epistemological ground only the power of magic. This power was working not by force pressing and challenging but rather by the principle of natural vacuum. Therefore, to search of magic on the ground of modern science cannot be a feasible idea but it would be more appropriate to say that natural processes had "magical effect."

Since a clear distinction between natural sciences and occult sciences was not made at that time, occult forces seemed to be responsible from the events in the phenomenal world. This perspective was predominant almost all scholastic philosophers. However, thanks to the perspective of natural magic, the idea that ordinary people too could have knowledge on occult forces appeared. To use some occult values or principles was among main applications of the Medieval experimentalists and alchemists.[32]

Roger Bacon adopted a scientific experiment methodology, even though he accepted the existence of occult forces. In this context, he stated the following in the discussion of speculative alchemy in his book *Opus Tertium*:

> But there is another alchemy, operative and practical, which teaches how to make the noble metals and colours and many other things better and more abundantly by art than they are made in nature. And science of this kind is greater than all those preceding because it produces greater utilities. For not only can it yield wealth and very many other things for the public welfare but it also teaches how to discover such things as are capable of prolonging human life for much longer periods than can be accomplished by nature.[33]

As it can be understood from Bacon's statements above, he clearly thought mechanical equipment and tools used in practical life within the framework of *scientia experimentalis*, which he himself developed. Bacon rejected their magical and mythological origins by stating that he remained within the scope of probability of these equipment and tools. Bacon was able to take such close care of the mechanical arts was possible thanks to a technician, called Pierre de Maricourt.[34] Even though various Archimedean mechanical systems continued, these were largely lowest level techniques necessary to keep daily life going. As for Bacon's framework, the human perspective, as the expectations revealed too, was not based on imitation but on *scientia experimentalis*.[35]

In addition, concerning the distinction between legitimate and natural magic, Guillaume d'Auvergne (1190–1249), known as a contemporary Avicennan philosopher, discussed natural magic. Remaining loyal to the St. Augustinian terminology, d'Auvergne maintained that

Magia Naturalis was of the human soul, which was tainted by the original sin and tended continuously to commit sins.[36]

Some arguments proposed in the field of the history of science since last quarter century claims that the question of magic stood on an important axle in the scientific revolution. To ground this claim, these perspectives tries to show that the idea of the classical magic found space in early modern texts. The critical point here, however, is why the magical framework was used. It can be observed that as magic developed an authority against the scholastic order by using natural forces in order to have an ontological legitimacy, the principles of the philosophy of nature grew to a level equivalent to it. In fact, many instances that the church objected as magic were actually natural and rational events.

2.2 "Privileged Knowledge" as a Question of Property

One of the most important parameters of the modernity related our subject was the solution it brought to the question of property. As it is known, the commercial power of the aristocratic class, which appeared after the mercantilism, became the starting point for an important transformation against the Church's monopoly of property. The rise of intellectual enterprises and artistic skills as a question of property and the synchronic resolution of this question became an important component of the new knowledge system. The difference of this new synthesis from the ancient systems was that it brought security and privilege to the management of hitherto unseen individualistically profit-oriented production.

The interest by both societies and political powers on handicrafts and original works has been great since the ancient period. In all political formations, whether small or large, people with such skills received encouragement and rewards for their labors. In traditional structures, any kind of power accumulation could be based on such skills but sustainable systems could not be established. The system usually has an innovative structure proceeding on that individual enterprises found previous actions and works insufficient.

Besides, master-apprentice relations played a critical role in the development of handicrafts. Keeping the knowledge on a certain production technique secret was important for the classical understanding of the question of property. However, these kinds of knowledge could not be kept eternally secret or privileged. Although techniques on various arts remained on the control of certain group by transmitting them from father to son or from master to apprentice or within guild system, they always had some kind of semi-transparent character. But the ultimate objective of these organizations was applying some social demands rather than controlling technical leadership or privilege.

Taking the example of alchemy, it can be said "in pre-Modern period, knowledge was not something to be owned but rather the knowledge itself owned human being." Due to this assumption, all alchemic traditions had been built on the idea that once certain secret knowledge was attained, human being would undergo certain spiritual transformation too. In this case, the relation with knowledge was far above and beyond simple property relations. Unlike we imagine today, almost all alchemists were not interested only with the practical applications of the knowledge they attained. This kind of relation, in Medieval Europe, led to the emergence of a world of esoteric knowledge that was far from reality and inaccessibly mysterious.

The whole classical tradition was built on the works of science and arts that treated past legacy with due respect and viewed its own novelty only an additional nuance. Taking advantage by usurping the idea or work of arts belonging to another person was considered ethically negative. The new understanding of property gained a character that cannot be limited by respecting the masters only through citing.

On the other hand, the rising organizations related to handicrafts in Europe sought to produce strategies toward keeping the technical know-how concerning these handicrafts within the limits of cities. The fact that the unauthorized leaving of glassmaking masters in Venetian Republic would result in punishment or annulment of license shows how the city-state administration controlled all the artistic production. The glassmaking masters petitioned for aggravation of punishment because sometimes occupational secrets were transferred to neighboring cities due to the failure of punishments' deterrence.[37] The governments of city-states, occasionally if not so often, announced amnesties, recalled the artists who fled to neighboring cities, returned their licenses and sought to maintain their monopolies in that field of art.[38] This kind of privileges forming an institutional basis of the patent system was prevalent in Venice.[39]

Occupational secret knowledge was critical for the survival of that city-state and sine qua non for the competitive advantage in the region. Generally speaking, like in most of Medieval units, this kind of licenses remained within families but did not hold any element encoring novelties. However, this idea of intellectual property played a highly critical role especially in the rise of patent system.

In the ensuing centuries, the apparent supports to all arts and occupations that were significant for the economy of city of Venice continued by granting patents. Franciscus Petri coming from Rhodes to Venice petitioned for a patent for a weaving loom and the council granted him the monopoly and ten year of protection for establishing the loom because it foresaw Petri's technique's vital contribution to the city's economy. According to this, anybody including the close relatives of the person would add nothing to this loom.[40] The administrative and legislative authorities did not pay attention to whether these petitions of novelty or patent were personally done or not. Although it seemed to be contrary to the interest of the local occupational units in the short term, these privileges led to the emergence of different dynamics for novelties in the medium term.

From the beginning of the fifteenth century, a distinction between copyright and non-material property was made and writing something and having the privilege to produce that thing were separated.[41] By the rise and spread of the printing press, an important development in respect to copyrights appeared; in fact, selling a book in multiple print editions could be done only with the print house with which the author had contract, and random printing of books would be as much as possible prevented.

The Venetian Council issued the first patent law in 1474. The law was justified by that the city of Venice would become a center of attraction for skilled and talented people of different origins.[42] When anyone among them developed an invention would apply to *Provveditori di Comun* and receive ten year of protection and within this period nobody of Venetian origin would produce that tool. Otherwise, on official petitions of complaint, the author or inventor would receive a penalty of 100 ducats and the thing produced would be annihilated.[43]

Early examples of patent among Italian city-states were also seen in Florence. Filippo received three years of protection for the new ship he invented for transporting loads in Arno River in 1421. According to this protection, he would be exempted from taxation, other ships would not make transportation on the river without the inventor's permission. Anyone who did not conform to this rule would be punished and his ship would be burned.[44]

The patent system that gained a standard level in Italy was transferred to many European cities by the migrating Venetian merchants and artisans. In various European locations, such as Antwerp, France and England, the Italians taught the occupational secrets to local apprentices and spread their arts through the help of similar institutions. The Venetian glassmaking masters not only transmitted their techniques but also contributed to the spread of the idea of patent.[45]

For about 150 years after the Venetian patent law, it became institutionalized with various nuances in all Western European countries.

Even if it did not completely protect today's intellectual copyright, these institutions were critical for the inventor's maximum profit from his own works and for the imports of *know-how* from neighboring competitive countries. Starting with the seventeenth century, in addition to artists and merchants, the scientists too began to contribute to the process. It was the class of scientists that systematically contributed, directed and reproduced epistemologically the artists' products.[46]

While these privileges were applied in certain cities and countries in a limited scale during the fourteenth and fifteenth centuries, the number of cities increased dramatically throughout the sixteenth century. Thanks to the patent laws, the inventors could prevent somehow others from abusing or stealing of knowledge and technique they discovered. Besides, the field of intellectual property was kept quite wide. The authors registered the ideas to their names by printing their works in book format in a print house and guaranteed the credit they might acquire from these ideas. Similar privileges were also true for the print houses issuing books. John Speyer, a German in Venice, obtained the right to print and sell all kinds of books by a decision issued by the city council in 1469.[47] Because of this privilege, nobody would sell any books brought from abroad. Even if the period of protection was five years, when the right holder died earlier the privilege could not be transferred to anyone else. His family who continued the occupation was not allowed to keep even remaining part of the privilege, but a note was added to the register book, "The right holder died, it is void."[48]

Another aspect of the mercantilist protectionist laws and applications was its conflict with the existing guild system. The new merchant class could concentrate the surplus value around their own profits by keeping the new knowledge they brought from long distance out of the existing production order. Even if this system was not claimed to be the ideal solution for the total benefit and social benefit, it was a foundational factor for the Western transformation, especially for the property rights. In addition, occupational representation and production rights of certain arts were given to certain people as monopoly for a limited time, which created a center of attraction for the methods of technical arts in neighboring countries and cities.

350 year after the privilege knowledge system was established, John Locke's theories of property was built on such a past background. While the individual obtained equal benefit rights on land and its products, the fact that the labor used for appropriating these products was shown as the origin of property—just as the colonialism was a similar appropriation—ensured the continuity of the mercantilist system.[49] Locke's philosophy on individual rights and property aimed at

keeping sustainable the existing hierarchical order of the state, people and bourgeoisie.

Notes

1 English Cardinal Kilwardby who had been influenced by the translation movements examined mechanical arts and verbal arts within human sciences, which was critical for establishing this relation. See. Paul Thom and Henrik Lagerlund (ed.), *A Companion to the Philosophy of Robert Kilwardby*, Leiden: Brill Publications, 2003, p. 351.
2 This is used for indicating the natural flow of knowledge from more intensive environment to less intensive environment.
3 Houses with wide halls have only one bed and the rest of the family would sleep in shakedowns. The used furniture (Fr. *meubles*) composed of portable ware that were assembled on a corner at night and disassembled and piled up on a side at daytime. For daily life and private life in Medieval Europe see Jeffrey L. Singman, *Daily Life in Medieval Europe*, USA: Greenwood Publishing, 1999.
4 First translations on alchemy largely belonged to Hugo Sanctelliensis, a Spanish priest from the twelfth century. The first translation on alchemy was Balinas's *Kitāb Sirr al-Haliqa*. See. A.Y. al-Hassan (ed.), *Science and Technology in Islam*, Vol. IV, Paris: UNESCO Publishing, 2001, Part 1, p. 140.
5 Paul David, *The Historical Origions of "Open Science", An Essay on Patronage, Reputation and Common Agency Contracting in the Scientific Revolution*, Stanford: Stanford University, 2007, p. 9.
6 The shorter version was translated by Spanish John in Toledo and the longer one was translated by Philip of Tripoli in the thirteenth century. See William Eamon, *Science and the Secrets of Nature*, Princeton: Princeton University Press, 1994, p. 45.
7 David Lindberg, *Science in Middle Ages*, Chicago: The University of Chicago Press, 1978, p. 485.
8 Paola Zambelli, *White Magic, Black Magic in European Renaissance*, Leiden: Brill Publications, 2007, p. 48.
9 Eamon, *Science and Secrets of Nature*, p. 145.
10 Aristotle, *Nicomachean Ethics*, 2003, p. 333–5, 341.
11 Aristotle, *Posterior Analytics*, p. 71b10ff.
12 Aristotle, *Nicomachean Ethics*, 2003, p. 333–5, 341.
13 The distinction between fact (*quia*) and cause (*propter quid*) was prevalent in the whole classical period. Aristotle asserted that *quia* was a knowledge on the fact and could not be represented by reason whereas *propter quid* could produce scientific knowledge because it also included the cause. See George Ovitt, "The Status of Mechanical Sciences in Medieval of Learning", *Viator*, Brespol Publishers, Vol. 14, 1983, p. 103.
14 For distinction see Thomas Aquinas, *Commentary on Aristotle's Metaphysics*, John P. Rowan (trans.), Chicago: Regnery, 1961, Book I (A). In addition, as an example to the modern distinction, Pieper asserts that liberal arts address to "common good", practical arts to "common needs" and both of them lead to "useful results." See Josef Pieper, *Leisure: The Basis of Culture and the Philosophical Act*, Alexander Dru (trans.), San

Francisco: Ignatius Press, 2009, p. 78. Besides, according to Pieper, philosophy was the freest of the liberal arts. Ibid, p. 38.

15 George Ovitt, *Restoration of Perfection: Labor and Technology in Mediaeval Culture*, New Brunswick: Rutgers University Press, 1987, p. 119.

16 Heidi Gearhart, "Theophilus On Diverse Arts", Unpublished Ph.D. Thesis, The University of Michigan, 2010, p. 21.

17 Eamon, *Science and the Secrets of Nature*, p. 85.

18 Eamon, *Science and the Secrets of Nature*, p. 87. It means "Now the master is speaking..." The author's citing of the alchemy master is an indicator that the language of alchemy was life in daily arts.

19 Eamon, *Science and the Secrets of Nature*, p. 120.

20 Eamon, *Science and the Secrets of Nature*, p. 87.

21 *The Pirotechnia of Vannoccio Biringuccio: The Classic Sixteenth-Century Treatise on Metals and Metallurgy*, Cyril Stanley Smith (trans.), New York: The American Institute of Mining and Metallurgical Engineers, 1913, p. 14. The first metallurgy book in English, *A Discovery of Infinite Treasure*, was published by Gabriel Plattes in 1639.

22 *The Pirotechnia*, p. 15.

23 William Eamon, "From the Secrets of Nature to Public Knowledge: The Origins of the Concept of Openness in Science", *Minerva*, Vol. 23, 1985, p. 329.

24 Eamon, "From the Secrets of Nature to Public Knowledge", *Minerva*, p. 332.

25 Eamon, *Science and the Secrets of Nature*, p. 134.

26 Abraham Rees (ed.), *Universal Dictionary of Arts Sciences and Literature*, Vol. 1, Edinburgh: Longman, 1816, p. 413.

27 Eamon, *Science and Secrets of Nature*, p. 135.

28 Eamon, *Science and Secrets of Nature*, p. 137.

29 William Eamon, "From the Secrets of Nature to Public Knowledge: The Origins of the Concept of Openness in Science", in *Reappraisals of Scientific Revolution*, David Lindberg and Robert Westman (ed.), New York: Cambridge University Press, 1990, p. 335.

30 John Porta, *Natural Magick*, (trans.) Derek Price, South Dakota: Nuvision Publications, 2004, Preface to the Reader.

31 Zambelli, p. 5.

32 Pierre Hadot, *The Veil of Isis: An Essay on the History of the Idea of Nature*, Michael Chase (trans.), USA: President and Fellows of Harvard College, 2006, p. 68.

33 A.C. Crombie, *Augustine to Galileo*, Vol. 1, New York: Dover Publications, 1959, p. 37.

34 Hadot, p. 116.

35 Roger Bacon almost never cited the Greek mythology. Like the Islamic tradition, he identified philosophers with prophets and mythological gods with philosophers. See Andrew George Little (ed.), *Roger Bacon Essays*, Oxford: Oxford University Press, 1972, p. 137.

36 Zambelli, p. 47. Similarly, when Pico Mirandola (1463–1494) asserted that natural magic was nothing but the most powerful side of natural knowledge, and thereby the highest and most perfect side of the philosophy of nature, he meant to explain it as part of the philosophy of nature. See Zambelli, p. 132.

37 Pamela Long, "Invention, Authorship, "Intellectual Property," and the Origin of Patents: Notes toward a Conceptual History", *Technology and Culture*, Michigan: Society for the History of Technology, Vol. 32, No. 4, 1991, p. 30.

38 Long, "Invention, Authorship, "Intellectual Property," and the Origin of Patents", p. 31.

39 These can be counted as examples of privilege before patent: According to the law of production promulgated by the Venetian Council in 1297, sales were allowed only in spaces belonging to Giustizia and the ingredients could be protected as secret from other pharmacists. A Council decision in 1323 promised an individual the support materially for their windmill production and a grant up to 80 ducats for the *experimentia* done by the person. See. Long, "Invention, Authorship, "Intellectual Property," and the Origin of Patents", p. 33.

40 Sa Yu, Political Privilege, Legal Right or Public Policy Tool? A History of the Patent System, ATRIP Essay Competition, 2009, p. 9.

41 Yu, p. 4.

42 Yu, p. 4.

43 M. Frumkin, *Early History of Patents*, Richmond: Surrey Fine Art Press, 1947, p. 49.

44 Long, "Invention, Authorship, "Intellectual Property," and the Origin of Patents", p. 35.

45 Frumkin, p. 50–54.

46 C. Belfanti, "Between Mercantilism and Market: Privileges for Invention in Early Modern Europe", *Journal of Institutional Economics*, Vol. 2, No. 3, December 2006, p. 319–38.

47 Frank Prager, "History of Intellectual Property 1545–1787", *Journal of Patent Office Society*, Vol. 26, No. 11, Canada: Compiler Press, 1998, p. 711–760, http://www.compilerpress.ca/Library/Prager%20History%20of%20IP%20 1545-1787%20JPOS%201944.htm#1._Introduction (15 July 2012).

48 Prager, p. 716.

49 John Locke, *Second Treatise on Government*, Canada: Dover Publications, 1956, p. 17. 61.

3   Quantification

It is possible to call *quantification* as perhaps one of the most important stages leading to the scientific revolution. In other words, the critical process was a new understanding of the qualitative forms propounded by the classical *physis* in the newly developed mathematical language, which became shared language following translations. The quantification was the name for building a shared language—this was largely a mathematical language—that facilitated the new dialogue emerging with the scientific revolution.

3.1 New Environment, New Awareness and Space-Time Concentration

From the thirteenth century onward, the new interpretations of Aristotle's commentators had already started a new knowledge revolution; in fact, the classical Aristotelian interpretive system took into consideration the whole system rather than individual facts. However, thanks to the Arabic translations, this system was largely adjusted; the researchers arrived at the conclusion that the method of deduction would not address the problems, but "what individual facts showed" was more important. Therefore, the systemic walls that stood as a barrier to the knowledge obtained from phenomenal world were demolished. This new approach influencing countless fields from alchemy to geology, from astronomy to philosophy of nature, became the starting point for a process that triggered many factors. The most important factor distinguishing this process from others was its allowing for a new understanding of nature, which paved way for the individual to redefine himself in the universe and develop a new awareness.

The *awareness* was a fact that can, in essence, be limited by the physical universe, but cannot be subordinated to it. This realism foreseeing a conventional framework that was differentiated from the irrationally

DOI: 10.4324/9781003275756-4

defined universe and the physical world as well as proceeding completely on ordinary experiences offered a kind of knowledge that was both open to every individuals and sharable, and spread publicly. It can be said that the whole process of quantification was basically the emergence of such a fact. For example, the kind of investigations that all cats everywhere in the world would not chase after mice created a possibility of a heterogeneous reality. Even if the idea of a fixed order or cosmos continued to exist, the gate for possible worlds was opened thanks to the nominal reality and multiple experiences. The quantification was important for determining the principle of both constants and variables in this heterogeneous structure. This process revealed itself, one of the most perfect stages, by Newton's mathematization of the principles of the philosophy of nature.

The quantification was in some sense something that can start by human beings' determination of his/her coordinates in the universe. Determining such a starting or moving point was critical for defining relations in the external world and their location in respect to human beings. Human beings' location could start not only by the perception of geographic space on the earth but also by conflict resolution and multiplication of consistencies through rationalization of mental processes. As a conclusion of conflict resolution, that is, free experimental reconciliation on facts and events, thanks to the criteria of public control and consistency, it was discovered that the "common sense" was a natural judgment rather than external or relative definition and could appear through only common participation and sharing by people.

The location stage of the quantification process ended up in another transformation in respect to the perception of time. The idea of *ancient/eternal* made possible the perception of modern times framing both the present at that time and also partially the future. Therefore, it appeared that the utopia of ancient/eternal could be interpreted as a project of future. This could be realized only by explaining and locating the present, in other words by the real time-space perception. For example, the utopia of Francis Bacon could appear by this kind of time perception and was translated into a real language by Newton.

As the perception of theological past was formulated largely through certain Biblical events such as the Flood, the Creation and so on, the creation of a historical identity was not possible. For awareness, the existing locations should be tested and transformed into sharable outputs in an institutional level. Even for a daily time a big confusion existed in many parts of Europe. Even though Roger Bacon indicated the need for a calendar reform in his letter to the Pope, this confusion would last for a little long period of time. Despite the efforts of many philosopher-scientists such as Regiomontanus, Nicolas Cusa and Copernicus, the theological centers of their times continuously delayed all the demands of change in order to conserve their monopoly of determining time. But as long as the pressure of social processes and developments required a sharable and reconcilable perception of time, the Church's resistance on this subject would decline.[1]

Again another transformation related to awareness occurred in the perception of place and space. As it is known, there was a limited perception of space based on the classical Greek cosmology. Interbedded transparent heavenly globes were turning around by cross effects to each other and causing all kinds of movement and change. The outermost globe was carrying static stars, those inside were managing the planets, the sun and the moon. The movements of all these interbedded globes were perfectly and supremely circular movement in accordance with the classical cosmology. The main matter of the heavenly bodies was free from all defects and was made of ether, an element other than four elements. The peripheral formations struggling against the scholastic structure having a defensive position with the classical cosmology would naturally gain the upper hand through this new awareness. However, the new awareness was the first stimulator on the way leading to the Renaissance, and it was not unique to the Western world. The following examples are sufficient to show what kind of disorder this awareness removed.

In addition to the classical cosmology, information on existence and human beings in various regions began to be articulated from the thirteenth century onward. While discussing natural and cultural conditions of various regions of the world in his book titled *The Travels of Sir John Mandeville*, Mandeville mentioned a sea made of pebbles having oceanic waves and the existence of men with one foot in Ethiopia.[2] While talking about cotton plants in India, he stated that the plant had small lambs on its branches and extended branches to the ground in order to feed its lambs and even it was eating this fruit itself.[3] Even if this kind of fantastic accounts was critical for showing changing and relative existential levels, it is a continuation of an unnatural perception. This perception of cotton in Europe continued for centuries; even during the seventeenth century vegetable lamb was to be investigated in the travels to the East.[4] One of the best examples to show how much the relations with reality were deformed and how much irrationality was there was Sir Robert Moray's "mussel bird" report prepared for the Royal Society. Sir Moray argued that he traveled the northern shores of Scotland and observed himself that a bird whose beak, feathers and wings were perfect was coming out from every mussel shell. His report that received great interest was published in 1678 by Philosophical Transactions, the printing organ of the Royal Society (Figure 3.1).[5]

One of the notable parameters showing in what kind of world the individual viewed himself or what kind of world he had was maps. Maps having a narrative capacity more effective than books were a scheme created at the level of absolute reality observed by people as well as had a powerful legitimacy for showing the route of a travel and giving directions for the further world. The geographic directions were defined in respect to both theological and cosmological symbols of the time. Because the paradise, the sacred place, was on the direction of east, it was shown on the upper side on the maps, and the direction of north was shown toward east, because north could find a place only on the east. The western direction of the churches used to look toward markets and eastern direction toward the altar. Generally, because a T divided O circle, the maps of the time were called O-T maps (Figure 3.2).[6]

Figure 3.1 Vegetable Lamb; the anonymous image used by Sir John Mandeville in his own book. Mandeville J., *The Travels of Sir John Mandeville*, 14th century, retrieved from wikimedia.org.

Figure 3.2 Examples of O-T maps from the first printed version of Isidorus's *Etymologiae* [Kraus 13], Augsburg: 1472, retrieved from wikimedia.org.

For a long time on these maps, Jerusalem stood at the center, Nile and Don rivers were the T's upper arms and the Black Sea and Aegean Sea were the leg dividing the map from north-south direction. On the east-west direction, Europe and Africa were divided by the Mediterranean.[7] As the maps, however, were largely drawn through theological symbols—for example, the paradise was drawn on the far east, at the top of Asia as a mountain rising toward the Moon— they included esoteric objects in addition to other hierarchically equal objects. Animal-like and mythological drawings on the map displaying local particularities were completely unreal and irrational objects. As these non-geometrical maps did not present qualitatively satisfactory information, they did have a usable value for sailors. This fact meant to undermine the idea that human beings could spread heterogeneously on the Earth. Understanding the whole world through mythological symbols would make ambiguous and often hide the "other." Besides, as the limits of the world intersected with the paradise, it ruined the globe and made the idea of human consciousness of globe impossible (Figure 3.3).

Although the need for fine maps as an important element of spatial quantification appeared long time ago, their parts of 15 degrees had not been seen in astronomic tools until the sixteenth century. The idea of meridian and parallel, albeit offered in some works at that time, was not generally reflected on the maps. On these works, however, for going from one point to another point some directives by saying the angle in respect to the compass needle, that is, the north-south direction, were

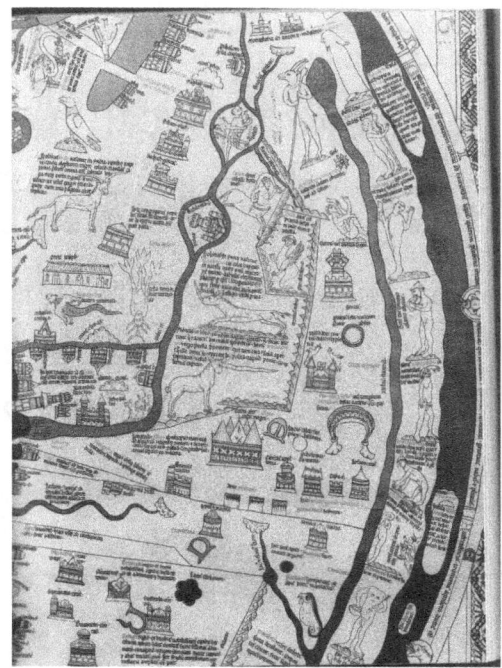

Figure 3.3 An example of non-geometrical map with mythological symbols. *Hereford Mappa Mundi*, displayed at the Hereford Cathedral, England, 13th century, retrieved from wikimedia.org.

given. Thus, one would go until the arrival point, in case of an error, one would return from the same route. Until the multiple reference system and new mathematical developments contributed to reading maps, old system had continued.

Both the irrational symbols and arrival points on the maps and the extraordinary existents in the literature showed that the ontological insecurity inhibited the formation of an existential hierarchy. This ambiguous and irrational structure that the individual produced for his external set informed us at the same time about his ontological position.

As mathematics did not generally have space at the level of fine transmission in the social life, certain words, like "some" or "a little," instead of words indicating quantitative certainty were used, for example in prescriptions, recipes, some handicrafts or medical formulas in the thirteenth century. The absence of mathematical symbolic

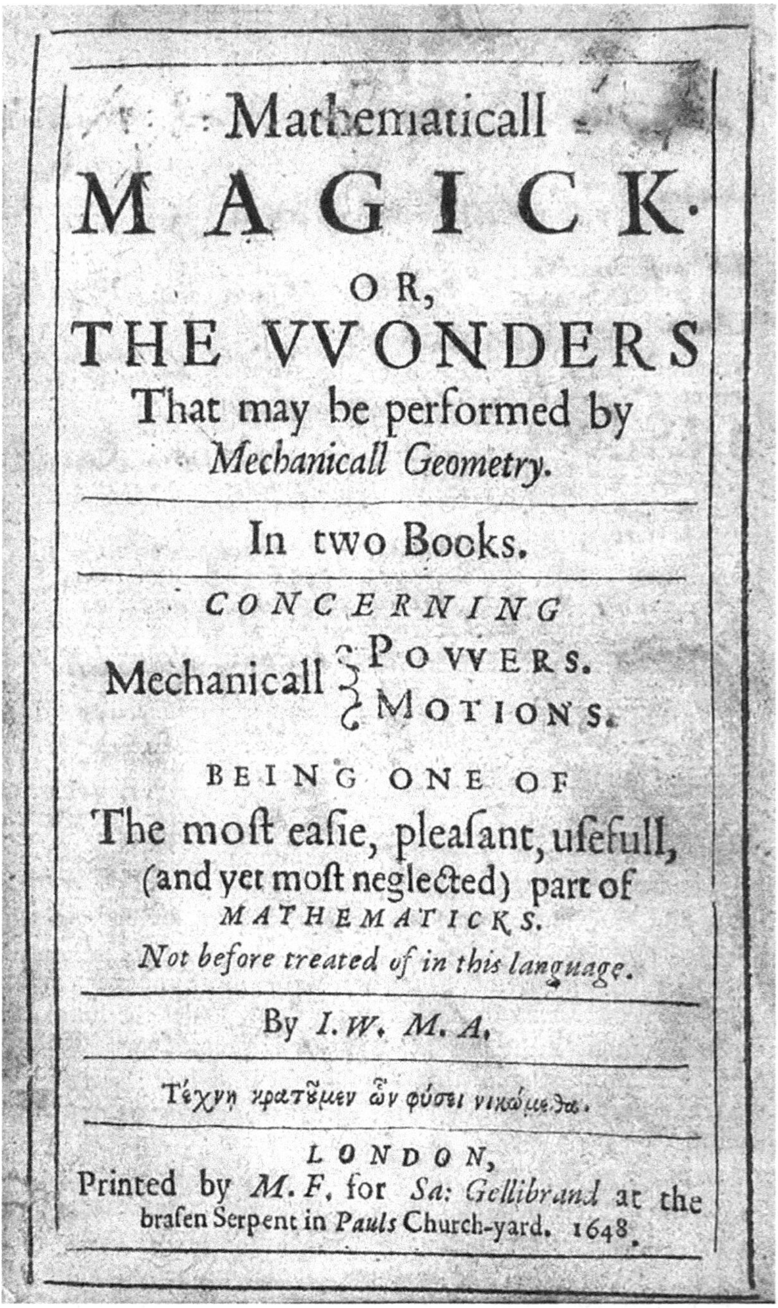

Mathematicall

MAGICK.

OR,

THE VVONDERS

That may be performed by
Mechanicall Geometry.

In two Books.

CONCERNING

Mathematicall } POVVERS.
} MOTIONS.

BEING ONE OF

The moſt eaſie, pleaſant, uſefull,
(and yet moſt neglected) part of
MATHEMATICKS.
Not before treated of in this language.

By *I. W. M. A.*

Τέχνη κρατεμεν ὢν φύσει νικώμεθα.

LONDON,

Printed by *M. F.* for *Sa: Gellibrand* at the
braſen Serpent in *Pauls* Church-yard. 1648.

Figure 3.4 Mathematical Magic, London: printed by M.F. for Sa. Gellibrand,
1648, retrieved from wikimedia.org.

language inhibited the calculation of the ratios of increasing or decreasing quantities/amounts or their representations. As the arithmetic quantities indicated by Roman numerals ordered side by side provided only limited information, they fell short of big numbers and multiple mathematical operations. The numbers used at that time in Europe did not fit to exact calculation techniques, and even the abacus method had not been known until it was introduced by Gelbert (Pope Sylvester II) in tenth century.[8] The spread of abacus would take a long time, and would become complete only by the development of accounting register systems in the fifteenth century. The whole Europe's relation to any quantitative value was based on these general assumptions before the quantification. Having mathematical arts at this superficial level became the harbinger of a critical role of the possible novelties in building an ontological security to the extent that they could be considered as a kind of magic (Figure 3.4).

3.2 Quantitative Quality and Anthropocentric Knowledge

The atmosphere dominating the Western world until the thirteenth century—especially in the cities, such as Toledo, Sicily, where the influence of Islam was felt—was of the character investigating and questioning what kind of knowledge would yield power. Many subjects such as Elixir, the Philosophers' stone, Talisman and so on were examined and investigated within the field of alchemy. However, it is possible to say that during the period following translations, a new quantitative physics distinct and free from Aristotelian physics rose. In relatively early period, Paul of Taranto stated in his book *Theoria et Practica* that human beings could make changes on the primary qualities either through arts directing nature or nature working under arts.[9] Paul separated "perfective art" for changing primary qualities and "purely artificial art" for changing secondary qualities.[10]

One of the important changes on the categorical distinctions in Ancient period was the transformation of the concept of *movement* and its analysis through a mathematical language. The quantification of the science of movement was introduced by Gerard of Brussels in the thirteenth century and developed by Thomas Bradwardine of Merton College in the fourteenth century. Following Gerard's studies, a group called "Oxford Calculators"[11] furthered the quantification studies. This group's studies seemed to prepare for Galileo's laws of motion. Bradwardine became the first person who algebraically showed the dependency of one variable on another one. His analysis mostly started with explaining the Aristotelian principles, especially the laws paying attention to objections in Arabic translations and commentaries. An object's movement was proportional to the size of the force affecting it and inversely proportional to the resistance of the environment. Aristotle was interested only in the way this object was to go within a certain time frame, but never did dynamic and kinematic analysis of this movement.[12] Concentrating on such kinematic analysis, Gerard sought to explain non-uniform movements with uniform movements by following Archimedean and Euclidean principles. He supposedly considered velocity an arithmetic concept like time and space. In the next step, Bradwardine made a clear metric definition of velocity; in other words, he specified that it was determined by the ratio of heterogeneous amounts to each other. In this, he assigned to proportioned amounts letters from the alphabet instead of numbers and mentioned verbally the division and multiplication of these amounts.[13]

In the ensuing stage, the quantification and the new motion physics began to evolve into two main streams; while mathematical kinematics gained prominence in Oxford, physical dynamics of motion developed in Paris. The common point for both was the creation of some quantitative definitions based on measurement, the use of concepts in mathematical methods and the concurrence of their perspectives on experimental measurement.[14]

Following the line of Bradwardine of Merton College, some names like William of Heytesbury (1313–1372/3) and John of Dumbleton (1310–1348) were interested in how quantities changed numerically in respect to a constant variable. In this context, "form" was a "quality" or "quantity" that could vary. The intension of this form was expressed through numerical amount ascribed or related to it. For example, one could talk of the change of heat as a variable form in respect to a constant form such as distance, time or the quantity of matter. Therefore, the uniformity of a velocity was to be explained by equal distance passed in equal time periods, or in the case of different lengths of way passed, it was to be explained as "accelerated non-uniform." According to William of Heytesbury, there were three categories of movement: distance (space), quantity and quality.[15] After defining the uniform movement, he explained "instantaneous velocity"; and defined "acceleration" as the velocity of velocity, "uniform velocity" as equal change of velocity in equal time periods.[16] Among all these examples, the most interesting one was that the distance an object with constant acceleration was equal to the average velocity within the same time period.[17]

As it can be seen in the explanations of velocity and acceleration, the relationship between quantity and quality reflected somehow in terminology. An accelerated velocity was considered to change instantly, and the instantaneous velocity was accepted as the intension or quality of velocity at that moment.[18] The Aristotelian thinkers asserted that the intension of qualities cannot be added up like distance or numbers. They could only arrive at a consistent outcome on what was changing in warm touching objects at the end of quantification.

Rejecting the classical categories, John of Dumbleton, who accepted the distinction between qualitative and quantitative, argued that the quantitative was the abstraction of reality. The rectangular coordinate system had been known by the cartographers and sailors since the ancient period. The *intension* of a quality showed the change on a normal line in respect to a certain *extension*.[19] The philosophers mentioned above were interested in general methods instead of certain physical facts. For example, Nicole Oresme (1320–1382) applied geometrical methods to the changes of sound and light in respect to distance.[20] He drew here the intensity or quality of the velocity as a normal line in every time unit, and showed its quantity by the area falling under the line connecting all these points—which expressed the total distance passed.

The reintroduction of these definitions to the scholastic system through Jesuits was important to show how Aristotelianism was transformed.[21]

In the context of mechanical sciences, while some concepts were quantified for theorical and mathematical usage, many Medieval mathematicians, who belonged to Archimedean-Aristotelian tradition of *Mechanica*, found remarkable—thanks to the translations from the Muslim world[22]—some subjects such as inclined plane, leverage, movement and specific weight. During the ensuing century, quantitative studies concerning physics of motion appeared in Oxford and Paris, two centers in Europe. Buridan's concept of *impetus* seemed to the first example for this development.[23]

More focus on practical arts by the *quadrivium* curriculum of the university or academia allowed for the training of high-skilled technicians.[24] During the fourteenth century, the measurement of time was standardized with the mechanical clock to 24 equal hours, and each hour to 60 minutes and each minutes to 60 seconds. This measurement was a quantitative time addressed to all and used commonly in place of varying time units of science and commerce. Similarly, the quantification of space would become possible by the mapping. This development was critical for a sailor to find his location on the map through the help of astrolabe and compass. In addition, after the discovery of Ptolemy's book *Geography* during the fifteenth century, location information would turn into the knowledge of latitude and meridian. For his maps, the starting meridian was Fortunate Isles, the farthest point known in the West.[25]

Similarly, images and qualitative subjects were quantified through geometry. Following the researches on the uses of lenses for magnifying and zooming the images, glasses were invented. French architect Villard de Honnecourt stated that like design and *portraiture* in building and machine making, geometry was also a powerful assistant contributing to the process.[26] The application of quantification in drawing and design began with the fact that, according to perspective principles, canvases were divided into grids like a chessboard. The thing enforcing the artist to draw so delicately was to show that the realist natural laws such as awareness and factuality were useful. The quantification evolved into the language that exactly enabled this. The ruled canvas was a tool that enforced and compelled artists to one thing, and conditioned and built necessarily a structure. This new form, through the quantification, allowed for a new perception of reality and its transmission as well as eliminated ambiguous and irrational definitions (Figures 3.5 and 3.6).

Music, which was a part of *quadrivium*, had its share in this quantification process and created a new language concerning how it could be repeated and transmitted. As it has been seen in many musicians'

Figure 3.5 Albrecht Dürer's perspective method, Germany: 1512–1525, retrieved from http://www.zeno.org—Contumax GmbH & Co. KG.

Figure 3.6 Albrecht Dürer's tool of drawing perspective, Germany: 1512–1525, retrieved from wikimedia.org.

works, theoretical novelties to explain relations between sounds began to spread from the fourteenth century onward.[27] In addition to the fact that all developments appeared in the field of practical arts through *quadrivium* in the classical classification, it resulted in the rise of the standardization of a qualitatively communicative language concerning physical world and facts, and of the comparable measurement. A particular practice could only be defined and expanded.

Ambiguous approaches to weight, an unavoidable part of the daily life, existed, too. Especially obtaining composite and mixed materials was related to the subject of weight. The studies on that certain materials could be obtained only through correctly determining certain amounts and repeating it were based in the fifteenth century. One of the firsts who commented on this, Nicolas Cusa (1401–1464), stated that changing ratios would create different composite materials; thus, determining correct ratios by experimentation paved the way for the connection between measurement and theory.[28]

The intellectual determination toward quantification, regardless of the developments in certain tools and mathematical methods, led to some changes in the habits of producing knowledge. Galileo, who has been indicated as the founder of the new science, and his followers did nothing beyond applying the Medieval concepts, such as *instantaneous*, uniform and accelerated movement, *Merton's rule of uniform acceleration*, into the physics of projectile motion and falling objects that they encountered in daily life.[29]

Similarly, measuring the weights and volumes of chemical quantities brought up a new science of matter. Weight had been identified since the ancient period as the only quality that can be measured outside of geometry despite its being a quality of object. The aim of expressing material qualities in quantity turned into defining other qualities that can be quantified through measurement by various tools. Therefore, Galileo objected to the Aristotelian classification based on opposites like heaviness-lightness and based his theory of physics on qualities, like weight, heat and so on, that could increase linearly on a single ruler just like the artists had been using. Following Santorio, who invented thermometer, Galileo measured the quality of heat objectively on a linear ruler. Later specific heat could be defined as a result of the scientists' and philosophers' distinction between heat and temperature. The quantification of sensible qualities like the methods used in practical arts and the development of quantitative tools would result in foundational changes in conceptualization, and particularly would lay ground for a new scientific project based on experiment.

3.3 Quantification and Functional Art

The tendency to find relations between the philosophy of nature and practical arts began immediately after the translations from the Muslim world into Latin. While the ancient philosophy of nature was based completely on explanatory or definitional consistency or assumptions/premises, some arts working in daily life such as measurement and toolmaking were working without relying on any postulate or assumption except for physical and material consistency. Pierre de Maricourt, whose mechanical experiences Roger Bacon made use of too, stated the following:

> While the investigator of this subject must understand nature ... he must also diligently use his own hands. He will be able in a short time to correct an error which he could not do in eternity by natural philosophy and mathematics alone, if he lacked care with his hands. For in hidden operations we greatly need manual industry, without which we can usually accomplish nothing perfectly. Yet there are many things subject to the rule of reason which we cannot investigate completely with the hand.[30]

The combination of the tendency to reduce philosophical theory into explainable forms and the tendency to obtain sensitive measurement and functions for practical arts resulted in the rise of "function-model."[31] The tools like globe telling how heavenly bodies had motion, clock and maps, albeit occasionally became lean, continued to be part of the Western thought since the ancient period. One of the examples coming to mind first was Honnecourt's work that contained drawings of a saw working by hydropower and of human bodies.[32]

During the following periods, thanks to the development of this kind of mechanical systems, mechanical clocks that was working with hydropower and regulating itself[33] based on weight and gravity began to appear. Besides, like many fields, this model became an effective form of display in optics. For example, Theodoric of Freiberg worked on a functional model showing how the rainbow formed with a magnified global raindrop.[34]

Mechanical settings in which a power and stimulant force worked synchronically and spread from a center stepped in to carry out complex functions. Crank and connecting rod mechanism was a combined system developed by Al-Jazari (1206) during the Medieval period. Here a rod that transmitted the movement by the gear mechanism magnifying the force and distributed even circular force to linear one was working synchronically (Figure 3.7). The printing press and various leverage systems were using this or similar complex mechanisms.[35]

All these numerous and complex applications in practical arts introduced a new method, rational functional modeling, which had not

Figure 3.7 An example from the drawings in al-Jazarī's *Kitāb al-Hiyal*, p. 137, 13th century, courtesy of the Süleymaniye Library.

hitherto been in the Western thought, into experimental science. The new relations with nature raised human beings to a different status of "knowing" and "making." The use of perspective in visual arts, mechanical clocks in time, mapping with trigonometric navigation and coordinate system in transportation as well as rational and functional model mechanics in construction and machine technique, systematic double registering in commerce and recording values in numbers were now the clearest examples of the beginning of a new theory-practice relation.

Another factor that contributed to this new theory-practice relation was the translations from Ancient Greek thought, especially from Plato and Hermetist tradition in addition to the Scholastic thought. The Greek atomism and Stoic texts were translated directly into Latin by Marsilio Ficino of Florence (1433–1499).[36] However, thanks to especially Arabic translations from Greek, a new knowledge/theory-practice relation alternative to the Aristotelian distinctions was defined and Plato was discovered under new circumstances. The establishment of a new metaphysics, therefore a new physics, without need to the non-functional Aristotelian distinctions continued for a few more centuries as being an original influence of the Islamic thought.

In addition to this framework, the whole ancient legacy would be reinterpreted by the rise of certain so-called scientific developments in the fields of arts; and all the Platonic texts, geographic texts, Euclid's geometry, Archimedean mechanics and the famous work *Mechanica* attributed to Aristotle and Vitruvius's *De Architectura* would be included into this process. By this new translation movement, the dominant Aristotelian influence in the first translation period was, in some sense, broken—as the translations from Arabic from the first period was already known, a new interpretation of mathematics became possible. All these developments were originated from the fact that practical and technical arts created social demands in daily and local conditions.[37]

Plato's rise to prominence during this period began by his portrayal as a superior philosopher than Aristotle by Petrarch (1304–1374) and by examining Platonism in a new perspective by Pletho (1355–1454). In his comparison between Aristotle and Plato, Pletho took Plato's side.[38] In all monotheistic religions, a kind of challenge between Plato's and Aristotle's cosmologies, on the one hand, and the creation theories, on the other, kept continuing. The most important for this challenge was taken by St. Augustine and Robert Grosseteste.[39]

An important Medieval thinker Mazzoni adopted the perspective of Timaios in his Aristotelian and Platonic comparisons in the context of philosophical authorities in theological texts, and held that primary qualities should be defined as matter's geometric or atomic forms and movements.[40] Pacioli (1445–1517) emphasized in his works practical mathematical arts and gave particular importance to mathematical and double-register accounting system, music, military techniques and mechanical and perspective drawing.[41]

From the fifteenth century onward, the rediscovery of the atomism of Democritus and Lucretius and printing Lucretius's *De Rerum Natura* (1473), Galen's works (1490, 1534), Sextus Empiricus's works (1562) facilitated the spread of Stoic philosophy of nature and Skepticism.[42] Influenced by this old but renewed trends, modern philosophers could produce a new kind of knowledge production. All the representatives of modern science from Copernicus to Galileo and Newton produced knowledge through these new trends.

Many translations from Greek to Latin, unlike the translations from Arabic four centuries before, allowed for new ways of solution—due to the Neo-Platonist and Stoic perceptions.[43] Therefore, instead of many wrong solutions of the Aristotelian cosmology and philosophy of nature, theoretical and scientific research based and built on the ancient models as well as focusing on more certain and correct solutions was encouraged. During the period from the fifteenth century to the early seventeenth century, a historical perspective against these philosophical approaches came into prominence through translations. Therefore, comparative histories for both Aristotelian and Platonic philosophy of nature and theoretical sciences could be written. The investigation of something in its historical process was important for showing what it actually represented. In other words, this new trend was actually a model matured by Islamic intellectual tradition. All systematic works of *Taṣnīf al-ʿulūm* would give a brief history in discussing any science.

Explaining nature through arts is considered to have started in the Western thought by depicting representation of natural existents and drawing various plants and animal after twelfth century. In fourteenth century, various landscape drawings in calendars, manuscripts and prayer books evolved into a different thing by the Italian techniques. Among the first representatives, painter Giotto began to reflect deeper views of human beings and buildings in his drawings.[44] When we examine the works before the thirteenth and fourteenth century that we can count as history of nature, we see that they generally used a geometric (symbolic or emblematic[45]) language. However, in the Muslim world at the same period, the classical representations based on mythological and Aristotelian designs were surpassed and especially the representations of plants and the animate were drawn in detail from

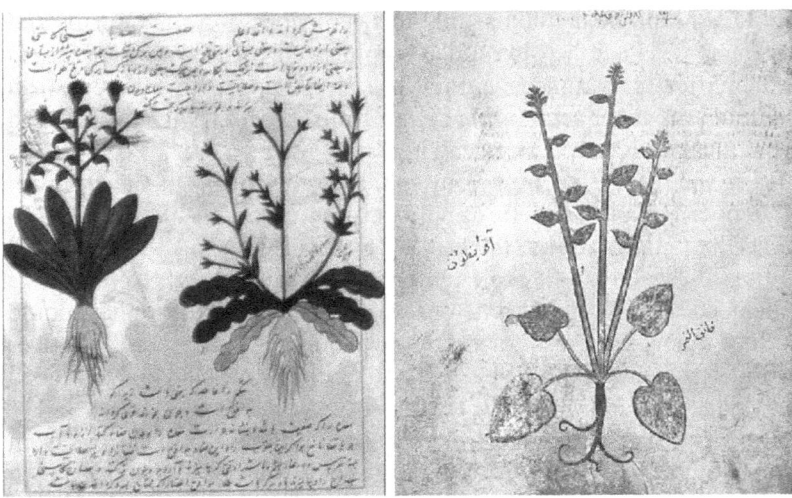

Figure 3.8 Examples from Muslim works on botany. (a), chicorium, illustrated in a schematic way in Dioscorides's *Materia Medica,* Topkapi Museum Library. (b), an illustration of the species "Akoniton napellus," folio 67V, retrieved from https://mybyzantine.wordpress.com.

their real views. (Figure 3.8).The realist drawings of human physiology starting from the sixteenth century showed a reference to human body to overcome an existential problem. Since the human being as the origin of the problem noticed that the solution was in himself, he began to take himself, that is, the individual, as the central focal point

As an example to observation, perception and transmission, one can count Frederick II's work *De Arte Venandi cum Avibus,* which he wrote after being influenced by the works translated into Latin by Arab translator.[46] Undoubtedly, since Frederick II transmitted the facts in the translated books through his own observations, he especially emphasized in the introduction "que sunt, sicut sunt," that is, representing things as they were.[47] In fact, the emperor used for himself the nickname "avis rassima," which means demanding insistently right observation.[48] He even stated that if no certain knowledge was obtained from observation, no conclusion could be drawn.[49] Applying the method he acquired in this work to many subjects, Frederick II revealed the fallacies on natural events in daily knowledge. This effort to "factual representation" matured in a long process from the middle of the thirteenth century to the end of the sixteenth century. This principle by its effects should be taken as a prerequisite for the modernity.

The theory and practice that allowed for the perspective discovered in the fifteenth century became possible by the *perspectiva naturalis* going back to Euclid as well as the technique that appeared with the use of many astronomic tools, navigation tools and maps.[50] Ptolemy, in his *Geographia*, showed latitudes and meridians while projecting the section that appeared by the representation of the Earth's global structure on a plane. Although this kind of works already existed in the Muslim world, they arrived in the Latin world through the Byzantine Empire only in the fifteenth century. From the first period onward, in addition to the translations that had great effects in the West, maps too began to be quantified. Even before its first print in 1477, *Geographia* affected the European mapping through the ways mentioned above—that is, translations from Arabic (Figure 3.9).

By the proportional representation of geographic locations on a plane, all oceans were discovered and the knowledge obtained began to be placed on the maps through available techniques. As the realist mapping technique foresaw mathematical reference, it contributed

Figure 3.9 al-Idrisī's world map rotated 180 degrees, retrieved from wikimedia. org.

to the development of perspective. In this sense, Filippo Brunelleschi, who has been considered to have discovered perspective, drew a cathedral in Florence on an art bit with a realist representation.[51] The artist drew the objects in this tableau proportionally in respect to their distance to a certain point and took as the reference point the point that united the horizon with the eternity. Scaling and the horizon in the eternity was actually a basic principle that should be observed in order to draw a picture in accordance with the fact.

Fostered by the humanist tradition, most of the artists in the fifteenth century thought that they actually sought something that had been told better in the ancient period. Unlike ancient philosophers, Medieval artists, by using mathematics and arithmetic, caught completely the principle of factual representation and made the language of art prevalent. In his work titled *I Commentarii*, Ghiberti (1358–1475) stated that no perfect art was possible without geometry and arithmetic,[52] and that following the methods and objectives of the ancient philosophers would bring perfectness to art.[53] Emphasizing his proportional and metrical observation of nature, Ghiberti stated that he tried to understand how nature revealed itself, how the views appeared and how they were perceived.[54] The reason that led Ghiberti and others to observe nature mathematically and quantitatively was that the existing perception of nature was drawn irrationally and thereby contrary to the facts.

In his discussion of necessary auxiliary sciences that a good artist should know at the end of his work, Ghiberti mentioned *perspectiva naturalis* and *perspectiva artificialis* and explained them through the anatomies of Averroes and Avicenna and the body measurements of Vitruvius.[55] Ghiberti asserted that it would be impossible to understand how to see without knowing eye's mold and stated that Avicenna and Ibn Haytham were principal sources for his book *Prospettiva* on this subject and that the thoughts of ancient philosophers were transmitted incorrectly in translations.[56]

Besides, Alberti (1404–1472) in his book *De Pictura* stated that he based his perspective technique on Euclidean visual pyramid, thereby explained it by the physics of seeing. One of the transmitters of this widespread opinion, Ghiberti proved it physiologically through examples taken from Roger Bacon and Witelo, two translators and commentators of Ibn Haytham.[57] Ibn Haytham, who maintained that seeing would happen by the perception by eyes and the distance adjustment by mind in brain, argued that the objects' distances could be found by comparing to other objects.[58] Similarly, emphasizing that distances could not be perfectly comprehended without a knowledge of reference, Ibn Haytham stated that the height of clouds, for example, could not be known or someone looking through a hole could not perceive the distance between two walls, being one of which taller than the other.[59] He argued that human beings under normal circumstances would take his own body, hands and feet, and would know the distance/height of other objects and formulate in the mind the height or distance of the distant objects by comparison.[60]

As Alberti approached to the art of drawing through the fact of seeing as a philosopher of nature, he described the perfect art according to the laws of physics. This explanation was actually an effort to form a language of description that would be checked, fitting to the fact and thereby corresponding to the nature of the object in the external world. Since every drawing was a plane section of a conical structure starting from eyes to the object's edges, artists should optically work on geometry and build principally the relations between the person and the external world.[61] Alberti, in his detailed description of how to draw an object, indicated that the object he took as a rectangular section should be conceived as a window looking to the world of objects, and that the artist could create the drawing from the dots on the lines that would start from the artist himself.[62]

Like in other arts, these novelties in visual arts focused on mathematical harmony and perfection, and showed, in some sense, that the art should procure a rational homogeneity in the accord between the object's parts and its whole. Just as the proportions of human body organs to each other was almost constant, proportional harmony would allow to establish a similar communication. Alberti defined the artist who used effective techniques by keeping this proportional harmony and was confident on himself as "virtue man" (*virtuoso*).[63] An artist continuously was to check himself due to "virtue" and aim at having this kind of a relation with his environment. As this definition combined the theory and the practice within the framework an *idea*, it implied that the artist actually tested or transmitted his ontological security over the art itself and at the same time it was important to turn rationality into a value or idea for an artist. The conceptualization of *virtue* man, *virtuoso*, was an important indicator in the modern Western paradigm as a prototype who could control himself and steer his thoughts. The perception of *virtue* foresaw that the human being was considered to be social and moral regulator in nature. This understanding developed the determinator character on himself and on nature by departing from the consistencies in relations among human beings or in nature. In its widest sense, *virtuoso* was a rational artist. Before constructing something practically with his hand, he first mentally test, build and approve it.

Alberti asserted that all machines would move like animals as if having powerful hands and lift weights like human beings' use of arms. Alberti, who said, "We should therefore imitate machines (*machinae*) like our contracting and relaxing organs," indicated that artist should be trusted more than many ancient philosophers just like some philosophers had been doing since the thirteen century.[64] The most important reason was the skills of technicians and artists to produce "consistent and reproducible" solutions to special or general problems faced in the daily life.

The fifteenth-century intellectual world reached a quantitative view of space through reinterpretation of the ancient geography and geometry. The most critical characteristic of the philosophers in this period was their reproduction of scientific accumulation in vernacular languages, which was the most important component that made all these quantification processes meaningful and provided knowledge with a new ontology. The aforementioned philosophers took to the next stage the transmission of the works translated into Latin by the generation of Roger Bacon in a consistent form and rendered a different structure, testing again the factual accordance with a new language within the restrictive-directive framework of the period's sociocultural conditions. Despite the presence of thinkers defending the monopoly of Latin as a scientific language, local demands and vernacular languages formed the dynamics of reproducible and consistent knowledge.

A vernacular language, Tuscan evolved into the Italian language of arts and literature. The Tuscan court, in other words, local centers of patronization, places a decisive role in this.[65] This was a development that weakened the controlling power of the scholastic circle who had retained the monopoly of producing and spreading knowledge. The classical texts were now, in some sense, criticized in vernacular language that the church did not use and this fact allowed for the opening of new channels. Besides, the fact that the new technicians did not know Latin had important outcomes in respect to the modernity and human mobility. The pressure of vernacular languages lasted until the beginning of geographic discoveries in the fifteenth century. The fact that the number of technicians knowing Latin in the port cities were insufficient and sailors were uneducated people entailed the production of naval maps in vernacular languages. The fact that the educational institutions where instruction was in Latin failed to respond existing social change resulted in that vernacular languages produced technical solutions. The Protestant movement could be seen as an extension of this in the field of religion. Among the factors inducing nationalization in Europe, mechanical arts having a chance to develop in vernacular languages should be mentioned. The principal item that removes the local-universal dichotomy was mechanical arts that allowed for the ontological transformation through the philosophy of nature and stood as a new imitation. Therefore, the individual overcame the predicament with a challenge and thereby a new ontological security was instituted.

The studies on ciphering techniques as a high-level form of knowledge in respect to the bases of modern science have important hints for our discussion. For example, Alberti stated that cipher (*formula*) should be based on a regular basis, and thus secret communication would become possible.[66] The demand for cipher was important for two reasons: first, the knowledge that had previously been considered by nature ciphered or secret revolving around alchemy gave its place to the ordinary knowledge that could be mathematically ciphered. Second, the human being himself could now hide the knowledge that represented by imitation the secrets of nature.

Therefore, *virtue* can be described as something completing or bringing up human being as a whole. One can look at the Renaissance and all the ensuing processes as a part of the human being's effort to define and place himself. This was for the artists an abstraction aiming at interpreting facts in a perspective sense and explain them in a form that could be approved publicly. The artists later took this technique to a further level. The reflection of the rules of scaling on arts allowed for building a technical language of display, and even one could see that these rules would be frequently used in the drawings of animals, plants as well as inside and outside of machines. Especially, Dürer developed several draft drawings showing the views of a machine or parts from certain angles and internal sections from various planes and their mutual modal relations.[67] Here the part-whole relation was the main understanding of quantified knowledge that appeared out of the positions of the parts in respect to the whole.

While almost all of the studies on the history of nature from the previous century had an "emblematic" perspective, the sociocultural practical applications and transmissions began the period of representation fitting to the fact. As the emblematic perspective allowed for the existing dominant structures to have subjective interventions, all visual novelties aimed at bare facts. Drawing was the field in which the most important outcomes of this new language appeared. Because there was a perceptive problem in respect to the reality, only a mathematical model that provided a natural agreement and carried the local perspective to the universal stage could allow a new language of communication. Since recounting or showing things as they were was the only way out for the local, local solutions, particularly languages, were the most basic dynamic of this process. Similar trends reflected also in other fields.[68]

All these developments in techniques of drawing and design showed their effects in architecture and engineering as well as visual arts. This new language was adopted by many architects and technicians, and was represented in their works because it provided the most efficient results. Da Vinci's works, Agricola's work on metallurgy and mines that contained the drawings of water pumps and mills,[69] and the detailed mechanical setting in the works of men like Agostino Ramelli (1588) were transmitted from one work to another, and they became reference books for many architect-engineers.[70]

Thanks to the establishment of this kind of perspective-based drawing techniques in the minds, a kind of visual interpretation based on rational deciphering appeared, which allowed Galileo to see mountains and oceans instead of weird shapes when he looked at the moon by telescope. It is reported that when the Jesuits took Ramelli's book on mechanical settings to China, the Chinese understood nothing, because the perspective-based visualization that accepted the thing looked was exactly the thing seen included adjustments according to certain geometric assumptions.[71] In fact, the perspective-based visualization technique, which foresaw the factual representation, included adjustments in respect to certain geometric assumptions.

As an indicator of the new period, Vitruvius's *De Architectura* carried the reflections of this new mathematical thinking arising in whole Europe. Vitruvius emphasized that no artist could advance by his own skills, he should take theoretical education; the knowledge of architect should be the children of theory and practice[72]; therefore, this architect should be knowledgeable in philosophy, history, physics and medicine for explaining the nature things as well as optics, mechanics, geometry and astronomy.[73]

Vitruvius in his book stated, "A machine is a combination of timbers fastened together."[74] While machines meant to realize a certain objective by many human labor, motor-automats (engine) meant to realize a certain objective by a single human labor, like *scorpio* and *anisocycli*.[75] Vitruvius continued by saying,

All machinery is derived from nature, and is founded on the teaching and instruction of the revolution of the firmament. Let us but consider the connected revolutions of the sun, the moon, and the five planets, without the revolution of which, due to mechanism, we should not have had the alternation of day and night, nor the ripening of fruits. Thus, when our ancestors had seen that this was so, they took their models from nature, and by imitating them were led on by divine facts, until they perfected the contrivances which are so serviceable in our life. Some things, with a view to greater convenience, they worked out by means of machines and their revolutions, others by means of engines, and so, whatever they found to be useful for investigations, for the arts, and for established practices, they took care to improve step by step on scientific principles.[76]

By the translation of both the works of Vitruvius and similar ancient texts from Latin to vernacular languages, some concepts like *machinate* or *mechane* began to be translated as regards their practical uses. The developments in drawing techniques contributed to show these works as more useful and thereby to raise their circulation. Since these kinds of translations were not compatible with the holistic scholastic system, mere oral explanations were not enough; they should be supported by drawings.

According to the quantitative thought's foresight, the mathematical reality would never find a place among natural things, because the things constructed and produced cannot be realized perfectly as they were in human minds. Therefore, mathematical arguments that had sharp certainty should be verified by physics, which had more certainty. Tartaglia made here a distinction between mathematics itself and the experience obtained by using mathematics and stated that if a theoretical argument was constructed completely, in other words, based on mathematics, it could have a reflection in physical reality. This distinction was an important stage for the "ideal" perspective, which disregarded losses and external factors in physical sciences.

Many neo-Platonic philosophers in the fifteenth century tried to look mathematics from the point Plato had looked and viewed it necessary in respect to education, dignity, virtue and so on, whereas many professional experts and artist working by mathematics preferred to look it rather through practical and vital perspective. Many Euclid commentaries belonging to the Stoic period were reprinted again throughout the ensuing century. These commentaries, unlike previous works, contained systematic information how they could be used by artists. Thanks to these translations, a theoretical outlook for the problems concerning nature emerged in connection to the applications of visual arts and mechanical arts.[77] The Greek skepticism's nature inhibiting to form definitive judgment led to rebuild the language of scientific rhetoric in fifteenth-century Europe. Throughout the century many universities slowly dropped almost all scholastic methods of learning and the Aristotelian logic processes for Epicurean, Ciceronian and Neo-Skeptic texts.[78]

The intellectual agenda of the time focused on certainty and the contribution of mathematical arts on this subject. Mathematic-physics relations as a debate since the antiquity continued again over Aristotelian and Platonic arguments. For example, Christoph Clavius, an important figure of the period, in his commentary on Euclid in 1574, transmitted mathematical sciences in an Aristotelian perspective in his discussion of logical differences between geometric postulates and syllogism.[79] Because mathematical sciences were both inherent in the physical world and imperceptible, he defended the position assuming that mathematical sciences were in between metaphysics and natural sciences. Just as the metaphysical matter was both mentally and externally independent and the physical existent was both mentally and externally connected to matter, the mind as a subject of mathematics was independent from matter, but externally connected to sensible matter.[80]

According to Clavius, none could understand representations and could claim to be an artist without knowing mathematics. Although geometry's origin was mechanic, it could not be limited only to lower mechanical tools, because geometry would procure our building a world by flying human being to the heavenly position he deserved and our management from there.[81] Besides, Clavius asserted that the curricular classes in Jesuit Collegio Romano that he himself organized should be taught according to mathematics, anyone who would be deprived of this would remain lame and deficient, and the rest of philosophy would be understood more correctly with the addition of mathematics.[82] Similarly, Guidobaldo del Monte (1545–1607) viewed in his work titled *Mechanicorum Liber* (1577) that mechanical arts were tools struggling against nature, making the things by imitation nature abstain from doing, and so having superiority over nature.[83]

The awareness of the relationship between mathematics and mechanics undoubtedly came from the need to transmit in a consistent way the roles mechanical arts played. Bernardino Baldi (1553–1617) shared the view that mechanics could be presented in mathematical way despite his description of it as a subject of physics. According to Baldi, mathematics was an absolute science before Archytas, but his combination of these mental principles with matter resulted in the capacity of making mechanics or machine; as for Archimedes, he combined perfectly geometry and mechanics' principles.[84] Similarly, Henri Monantheuil, a student of Ramus (d. 1572), asserted that while creating nature God not only geometrized but also mechanized it (*Deus geometricus et mechanicus*).[85] In the introduction he wrote to a work, *Mechanica*, attributed to Aristotle, he emphasized that mechanics was the acutest and most powerful of the arts.[86]

3.4 Epistemological Effects of the Quantification

The quantification had generally two footings: the theoretical one was processed by mathematics and the practical one with measurement. The extraordinary developments in mathematical arts whose clear effects were seen from the fifteenth century onward began by the introduction of algebra into European literature through the merchants' accounting registers. Even those who kept this kind of accounting registers or those who designed geometrical tools were called "mathematicians." The application of this mathematical knowledge on astronomy, optics and mechanics produced reformist results. The inclusion of *algebra* allowed for these sciences and arts to be perceived and transmitted more easily.

Besides, the need for precise measurement in practical arts resulted in different interactions; the techniques of precise representation, measurement and registering in many professional fields such as painting, music, sculpture, military equipment, architecture and accounting. Especially from the sixteenth century onward, the developments in navigation and cartography and the arts in many fields such as, first, architecture, then mechanical clocks, navigation and astronomic tools, mechanical settings, spread primarily in Italy, then to whole Europe. The solution of vital problems by mathematical arts came into play as a factor relieving the questions of ontological security. Because of responding such a practical need, it became attractive to specialize in mathematics and each one of mechanical arts.

From the end of sixteenth century, for many philosophers, Aristotelian primary qualities turned into new concepts that could be quantified. Departing from their life experiences, many philosophers displayed different approaches what could be quantity and what could be quality out of ten centuries of data. For example, Della Porta, in his famous work *Magia Naturalis*, mentioned a tool with a glass bulb on the top to measure temperature.[87] A similar tool was mentioned in many works, too. As composed from ancient pneumatic and hydraulic works, the first example of a kind of measurement and display was developed by Bartolomeo Telioux in 1611. He explained it as that once plunged reversibly into water, the water within the tube on top of a glass globe would move when the water got warmer.[88] One of the first publications on this device was a medical book by Santorio (1561–1636). It asserted that the device would measure temperature everywhere and in every condition and even in human body and it was assisted by a scaled ruler.[89] Jesuit Giuseppe Biancani, in one of his works published in 1620, defined a similar device as *thermoscopium* and described it; in addition, he asserted that Santorio was its inventor.[90] As for Galileo, he mentioned that he became aware of the presence of a similar device through one of his friends in 1612 and that temperature could now be measured. Even Galileo claimed that he himself was the inventor of the device because he stood as the first person who explained its working principle.[91]

Thirty more years needed to pass in order to have Toricelli (1608–1647) and Pascal (1623–1662) understand that atmospheric pressure was somehow effective on thermometers and for precise measurement they should be tested independently. The thermometers produced with hermetically filled liquids, most perfectly by mercury, appeared only in the eighteenth century.

Therefore, the seventeenth century was a period in which many quantified qualities were defined and measuring devices were invented. The devices like telescope, microscope, barometer, mechanical clock and air pump played a very significant role in collecting information on the perceivable world. The beginning of this process was not collection and reproduction of information. Using these instruments, one would show how nature could be perceived or that transmission and reproduction of nature was procured through a commonsense principle. These instruments were, in some sense, artifacts produced by philosophers. However, philosophers' capacity to make this kind of instruments was not by the achievements they obtained through speculation on the philosophy of nature, but rather their transmission of techniques they learned from practitioners. The capacity of masters making daily tools such as clock or compass paved the way for philosophers to adopt similar practical methods. Because of the demand for masters of instruments, about 20 workshops were opened only in London in 1747.[92] As each instrument was working according to mathematical and physical principles, the principles of the philosophy of nature that a philosopher had prejudicially never approved now fulfilled the task of "continuous rectifier." From this stage onward, the plane of relations between technology and science began to be built on instruments. This relationship whose examples were limited in number at that time became stronger in time, but it would require to wait until the nineteenth century to have an institutional relationship.

Throughout the seventeenth century, the measurements on some phenomena that were the subjects of the philosophy of nature became possible. For example, we can count here the determination of the angles of light refraction and reflection in respect to environment, [93] the frequencies of sound vibrations, tests and experiments on falling bodies, works on physical characteristics of various materials like metals and wood, and the more precise determination of specific weights of many fluid and solid matters. It was an important indicator that many philosophers knew and used various devices of measurement concerning astronomy but no tool belonging to the Aristotelian philosophy of nature was used or discovered. Was the deepening of these works result of a continuing process or did it mean that a new thing was proposed?[94]

Looking at the fast process of change in the seventeenth century under the lights of these developments, Dalton's atomic theory, Harvey's blood stream, Gilbert's magnetism and Galileo's falling bodies and projectile motion summarize the arrived point. Eventually, the more advanced mathematical structures' creation of a systematic language in determining natural processes would reach the summit by Newton and Leibniz at the end of the seventeenth century, and arrive a new stage by the development of geometry after Euclid during the nineteenth century. In fact, the basic characteristics of different elements involving in the process of change was the production of knowledge that passed the quantitative tests concerning matter and nature. Almost all people involved in the process called "scientific revolution" were the last representatives before the appearance of natural science, because they were the individuals who cared to present all the principles they outlined also as the subject of the philosophy of nature. The way Renaissance and post-Renaissance artists perceived nature and the way the philosophers of nature did so had a common intellectual ground, that is, virtue.

Starting with the nineteenth century, all inquiries on nature began as systematic and theoretical knowledge. Subjects of physics were ordered as heat, magnetism and electric, and they were largely based on their connection to mathematical structures. The difference of these structures from the classical mathematical structures was not only rooted in whether a phenomenon fit to mathematical explanations but also in whether their subjects were built in a way that could be verified.

3.5 Galileo and Rebuilding of Scientific Demonstration

One of the most important representatives of the new science (*true scientia*) proposed new arguments and assumptions on the classical world's most basic distinction as sublunar and supralunar worlds. His satisfactory explanations on specific events formulated a language that can be applied to the whole field of nature. Galileo, with a support of the ancient thought he relied on, indicated that a language of nature could be built by saying that "[the book of nature] is written in the language of mathematics, and its characters are triangles, circles and the geometrical figures."[95] The intellectual struggle he had throughout his life was remarkable for showing in which context objections or solutions *true scientia* that he proposed brought up. His works, mostly censured by the Church *Index*, should be examined in respect to ontology-epistemology relations. Although the philosophy of nature had problems in many fields from mathematics to religious thought and theology, Galileo built, or needed to build, the knowledge system by using a new language that was not identical to any of them, the pure language of the philosophy of nature. The fact that he was patronized by the court as *Natural Philosopher*, the unique intellectual actor who could counter and challenge all the established authorities of knowledge shows the kind of investigation he was in.[96]

Unlike the theoretical approaches of the time, the new principle Galileo proposed was to produce holistically the knowledge of each one of the phenomena by taking the experiments and results into consideration.

We can outline the new knowledge order of the method that Galileo tried to develop under several headings: first, the competition among nation-states of continental Europe that deepened at the beginning of the seventeenth century held the system of patronization effectively. The fact that the centers with power of patronization reached a certain level of population concentration allowed for the system to spread. Many dynasties' involvement in this competition, even if shortly, indicated the beginning of a new dynamic period in respect to knowledge-power relations. Exactly at the middle of the process going to the order of Westphalia, Galileo was the representative of a period in which the patron powers provided all kinds of instruments against the present influence of the Church. As the support he received from the courts coincided to a period in which the balance of power was changing, his theory of motion mostly built the epistemological ground of the rising actors in the new order.

Second, Galileo's new physics of motion and two new sciences he defined in this field, dynamics and kinematics, [97] concentrated on analyzing the motion systems that developed in accordance with the period's social dynamism. Galileo's investigations stood important for showing the natural interactions among all highly mobilized social elements. A transition from an understanding of stagnant or recurring motion to another understanding that could be expressed with changing and surprising natural processes could be observed at the same time as a reflection of the chaotic situation or as an effort to find a solution. At the end of this process, the perspective of "motion classified" that legitimized the whole competitive structure and led up to solutions diluting the pressure was adopted. The tranquility necessary for social mobility required classification and distinction in natural and forced movements in the physical world. The dynamics of social movements had a multipolar structure in continental European context, and had a new structure based on classical borders or on "far point effect" and free from competition of double or of "powerful centers" (of competent elements whose centers did not approach). The vernacular languages distinct from Latin were harbingers of order based on nations. The reinterpretation of the classical literature of philosophy-science and arts in vernacular languages in addition to being outside of the official/central structures allowed for their reproduction generally free from old background at the conceptual level, thereby their reinterpretation in a sense according to conditions of competition and their being a new element of the competition. This process allowed, at the same time, for the narration of the natural process with individual observations and languages. Local conditions paved the way for free production by keeping general and central thinking styles—that is, the holistic system and authorities—at the secondary level.

Third, one can take the controllable knowledge system that Galileo tried to obtain as support to a self-perception in transformation that defended the present cosmology against the theological centers. The systematic critiques brought to the present cosmology starting with Copernicus were of the elements contributing the most to the process of constructing a new ontological identity. The present Church-centered worldview had a scholastic thinking structure based largely on the Aristotelian philosophy. The traditional self-perception based on this systematic were declining, in the face of the multi-structures, the competition between the local and the universal and underwent a great tension.

Fourth, mathematics and practical arts had remarkable characteristics in Galileo's knowledge system. Although Galileo did not himself invent certain instruments, his experience was an important example that introduced them into this movement and gave them meaning in this competition milieu. It is known that he taught mathematics at University of Padua and he himself produced some tools of measurement and sold them. The experiences of designing these tools and producing them mechanically allowed Galileo to observe certain theoretical concepts such as measurement, validity and reproducibility. These tools that began circulating especially from the fourteenth century onward were demanded in courts, universities and many similar commercial places and as their making required special expertise, they could not be imitated easily. The best example for these was Galilee's telescope. Although he was not the person who first invented telescope, as he understood its working mechanism correctly, he has been known the person who made the most sensitive telescope and his products were preferred by various centers.

Unlike the conventional use of the telescope by its inventors, Galileo's turning it toward the sky was, of course, partially because of his experience in practical use of measurement tools. It is possible to say that he discovered the telescope for a completely different and new objective thanks to his capability of making tools based on his experimental methods. As having a hypothesis on what a telescope turned to the sky could see and what it could not was absolutely necessary, the connection between the instrument and what was to be discovered became more important. The maps of the lunar surface Galileo himself drew gave the best examples for an effort to freedom from the emblematic perception in accordance with the principle of "factuality." The search for certainty and rationality in fine arts was based on similar causes for the optical results obtained from some devices like telescope. Telescope was an instrument that was not only intended for seeing the objects in the sky more closely but also for zooming the distant objects through lenses, for which aspect it had been known and used for five centuries (Figure 3.10)

Even if he used telescope differently from its classical objective, Galileo did not abandon the Aristotelian system;[98] but he even emphasized knowing the principles present in nature to turn the knowledge obtained from natural processes into scientific knowledge.[99] Besides, according to him, only correct statements would bring correct knowledge, the object of scientific demonstration should at the same time be not only a mental imagination but a real thing. Therefore, science was not for something not demonstrated such as space or infinity, because they are nothing.[100] As only the knowledge on the causes that have demonstrations in nature was real, the demonstrations from virtual causes could not yield knowledge without certain qualities—for they were *ex supposition*. If preliminaries were not both correct and self-proven, acknowledgment concerning correct knowledge would not occur.[101] Galileo's all these scholastic works were not based on mathematical demonstrations, but rather full of arguments based on syllogism and logic. As he tried to reach universal principles from certain assumptions, the identity and strategy of the new philosophy of nature

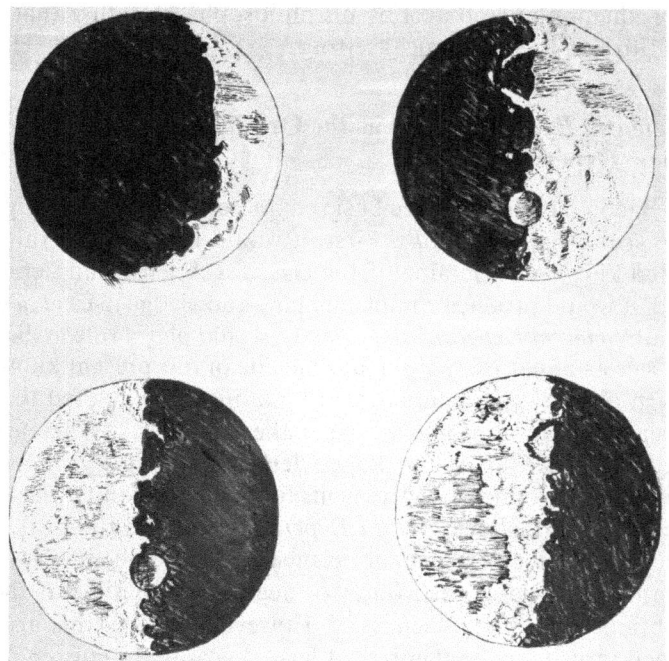

Figure 3.10 Galileo's drawings of Lunar surface. *Sidereus Nuncius*, published in March 1610, retrieved from wikimedia.org.

developed in a structure taking into consideration both holistic explanations and individual facts. That is why Galileo stated that whatever he would say would be only through examples of "what had been said" or "what had been proved mathematically," otherwise would not be possible, and he learned this method from the mathematicians.[102]

Galileo learned to turn the whole physical world into adaptive and consistent mechanical system from the atomists and Plato. Archimedes taught him in an analytical way geometric forms and mechanical settings as well as reduction to the ideal condition despite disorder, defects or problems such as friction because of material or process.[103]

In short, the distinguishing feature of Galileo's works was related to the character and importance of his audience. Starting his career in production of mathematical tools at the University of Padua, Galileo viewed the principles of the philosophy of nature as the most important element to form "true knowledge-science" thanks to his personal experience in instruments and used this as a starting point for establishing an ontological security. He could oppose the present centers of knowledge production and control by his philosophy of nature that he obtained through his mechanical settings.

3.6 *Scientia Experimentalis* as the Origins of New-Experiment

Roger Bacon mentioned three basic characteristics of his newly developed *Scientia Experimentalis*. First, it would first test the validity of deducted knowledge obtained from classical Aristotelian definitions. Second, it would produce complementary knowledge in case of deductions fail. *Scientia Experimentalis*, lastly, would play a role in discovery of the secrets of nature that remain outside of our present knowledge. We learn objects' characteristics and qualifications related to nature by experiments, for example, we can make machines.[104] Therefore, Bacon recommended his friend Pope Clement IV that *Scientia Experimentalis* could produce effective arms for the Christianity.[105]

According to Bacon, *Scientia Experimentalis* was the peak of our knowledge, master of all other sciences and had them work under its command. It would test, verify or negate by experiments the findings obtained by other sciences.[106] However, Bacon's this argument remained largely at the theoretical level, because an epistemological ground that could process such an experimental knowledge through control mechanisms did not exist at the time he was living in. In fact, the demarcations between knowledge and fallacy did not become clear at the sociocultural level. Another factor was that a serious insistence or institutional demand for such a pursuit did not exist. But the new tension that started with Bacon kept its pace and lasted until the end of the seventeenth century.

By the Renaissance, the experimental method seeking answers to universal questions posed by Platonic-Aristotelian worldview since the ancient period was adopted. By the seventeenth century, this method was replaced by the controlled experiment that appeared by a certain agreement between logic and art, generally took an Archimedean technique to center (*tekhne* [art, technics] or *ars*), making scientific investigation at technical level by using mathematical and mechanical ways and seeking answers to particular questions. In this case, the perspective of the research person began to be determinant—for it would create particular result for each experiment. At the same time, the environment and physical conditions in which the experiment was carried out were effective to change the results directly. In this sense, it became necessary to count the potentials and limitations of relevant technical equipment, measurement errors and approximate rounding as determining factors or accept them as part of control system. Therefore, the idea that every researches and investigations had special characteristics and thereby should be subjective turned into an accepted reality. When this fact was read inversely the thing sought appeared as an inductive system composed of individual experiences that everybody could reach by a shared rationality and feeling rather than deductions. The individual world of realities that everybody perceived one by one would turn into a shared accepted phenomenon. The thing verifying this fact was that it would first start by an individual demand, then would be put forward for the approval of the universal reason and ultimately would be expressed in a standard form that the majority would accept.

The limitations of this sort to the certain or correct knowledge in the past were perceived as epistemological limitations rather than impossibility to obtain absolutely certain knowledge on the changing world and they appeared as a threshold to pass over. Looking at the context, the fifteenth century was a period in which the historical background of real or functional representations of the technical and mechanical devices was discovered or reestablished. However, the remarkable outputs of the social memory appearing out of the practical knowledge or practical arts based on these devices could appear only toward the end of the next century (i.e., the sixteenth century). The accumulation of this energy became possible by the rise of accidentally increasing experiences and researches with *individual initiative* and *public approval*,[107] their translation into many languages and gaining more refined character by each translation.

This kind of quests for knowledge and experimental results concentrated primarily on the questions, concerning the scientific character of researches, of what the characteristics of the phenomenon natural objects were. Measuring the natural processes by various instruments, defining this knowledge in respect to a qualification and the logical connection between these processes were the basic elements that this new knowledge (*true scientia*) was based on.

True scientia brought to the fore remarkable novelties in practical arts throughout two centuries and began to challenge general assumptions on cosmology from the middle of the sixteenth century. The kind of knowledge that was tested dozens of time(s) everyday had a great potential for being a candidate for solving many problems of this sort. In this sense, it was possible to take Descartes's motto *cogito, ergo sum* as "I think differently, thus I exist." His most important work, *Discourse on the Method*, was in fact written as an introduction to a book on optics and mathematics.[108] *True scientia* became one of the most powerful and consistent ways to challenge and resist against the present ontological belonging system. Experimental demonstrations in various arts and sciences provided the cosmology a new testing setting that was based on sensitive measurements and consistent logical structures.

During the seventeenth century, the knowledge hunters of the new system applied reproducible and controllable knowledge form to the philosophy of nature and classified again the processes that became subjects to big philosophical systems. Therefore, they formed a new style of knowing for the universe by shifting the reference system privileging metaphysics-theology to physics.[109]

The main strategy of the representatives of the new knowledge system was to investigate and record the principles as individual phenomena that would yield certain experimental results rather than seeking to discover grand philosophical results that would be applicable in whole nature. Unlike other philosophical schools of their time, they bracketed ancient traditions if necessary and chased only after a style that could be "reproducible" and "modellable." The investigations from the skeptics until the early fifteenth century, on the other hand, were limited to what kind of knowledge could be possible and whether the obtained knowledge had absolute certainty or not. While discussing the possible kind of knowledge and whether this knowledge had certainty or not, phenomenal researches were left behind. The main subject of the classical philosophy, be it Aristotelian, Platonic or Hermeticism, was search for the essence of things and this composed mostly of answers to the question "why." However, this established order was challenged by the foresight of the obvious by the experimental and mathematical researches. Therefore, scientific research processes based on traditional logic and syllogism turned into a structure prioritizing the model based completely on testing and adopting the mathematical thinking as guide.

Besides, the philosophical and logical structure on which the new kind of knowledge was based brought successful solutions to practical problems even if it could not satisfy classical logical criteria which were necessary for real and certain knowledge. Mathematical arts developed a new knowledge system through *tekhne* and bypassed the existing *epistheme*. Those who were after the new knowledge maintained that to know something certainly with the new research methods they developed and mentally cooperated with could only be possible by making that thing with mind or hand. Therefore, artificial things or objects somehow surpassed the natural objects. The occupational formation of these people fitted to the combination of *natural philosopher* as they test nature by science. The occupational formation of natural philosophers was generally built on producing a kind of knowledge that was using mathematical tools and consistent on natural events and processes, reproducible and relatable as well as open to public scrutiny. Therefore, the experimental science evolved into "a collection of principles" composed of individually "observable, correctable and verifiable" decisions and disagreements.

The style of knowledge production on nature with mathematical methods that appeared with Skeptic and Platonic influences was an important starting point. The fact that the representatives of Neo-Platonism, such as Ficino and Mirandola, connected God's acts and human products and the efforts in the fields of fine arts and mathematical scaling contributed to a new kind of knowledge.[110] One of the examples that we can count for this kind of knowledge seeking was the introduction to an Euclid translation by John Dee (1527–1608). In this introduction, Dee examined mathematical sciences and arts by starting from structures in the style of Vitruvius.[111] He stated that mathematical arts were mathematical representation techniques and adjustable arts whose subject fitted, and a mechanician was a person who could make a job through senses without knowing mathematical representation. He asserted that speculative mechanician should be defined as the person who knew mathematical representation in addition to mechanical works.[112] In this work in which mathematical arts were classified from astronomy to music, from navigation to architecture, he especially recommended his readers Vitruvius's *Architectura*, Dürer's *Perspectiva* and Agrippa's *Occult Philosophia*.[113]

While the epistemological basis of Dee depended on Roger Bacon, we see that he adopted the approach that "an architect would first design first in his mind and then produce this with his hand," and mathematical arts were defined as "related to subjects not spoilable and with solid materials." According to this definition, architecture was part of mathematical arts and sensible materials and nature were its subject.[114] Although architect would direct mechanician for practical application, he himself generally would form demonstrations and reasons.[115] Again according to Dee, mathematician, someone like Archytas who could have his artificial goose fly, or a philosopher as follower of Heron could not be called as magician. This art was *archemastrie* that was offered in mathematical terms and produced as a real philosophy of nature and it included all experiences through helps of other arts.[116]

According to Vico (1668–1744), human being could only understand what he could make. Therefore, Latins used the words *verum* (true) and *factum* (made) interchangeably.[117] Starting from here, Vico offered a new metaphysics, and asserted that human being could not understand the basis of nature, but could only understand the products appearing through arts.

Notes

1 Gordon Moyer, "The Gregorian Calendar", *Scientific American*, Vol. 246, May 1982, p. 144–50.
2 John Mandeville, *Book of Marvels and Travels*, Oxford: Oxford University Press, 2012, p. 72.
3 Mandeville, p. 106.
4 Henry Lee, *The Vegetable Lamb of Tartary*, London: Sampson Low, Marston, Searle, & Rivington, 1887, p. 2. Lee states in his book that this vegetable lamb wasted dozens of years for the country's intellectuals.
5 Robert Moray, "A Relation Concerning Barnacles", *Philosophical Transactions of Royal Society*, London, 1677, Vol. 12, No. 137, p. 926.
6 William Cartwright and G. Gartner, *Cartography and Art*, Berlin: Springer, 2009, p. 149. O-T (*Orbis Terrarum*), that is, the abbreviations of "round" and "space"; it might sometimes contain textual explanation instead of geometric drawing with these letters.
7 W. Dilke, *Greek and Roman Maps*, Ithaca: Cornell University Press, 1985, p. 173. In addition see David Woodward, "Medieaval Mappeamundi", in *History of Cartography*, Chicago: University of Chicago Press, 1987, p. 340–1.
8 David E. Smith, *History of Mathematics*, New York: Dover Publication, 1958, p. 72
9 William Newman, "Technology and Alchemical Debate in Late Middle Ages", *Isis*, Vol. 80, No. 3, September 1989, p. 445.
10 The author defines the primary qualities as warm, cold, wet and dry, and the secondary qualities as black, white, sharp, blunt, bitter and sweet. See William Newman, *Promethean Ambitions*, Chicago: University of Chicago Press, 2005, p. 69–71.
11 *Oxford Calculators* was composed of four people: William Heytesbury, Thomas Bradwardine, John of Dumbleton and Richard Swineshead.
12 Walter R. Laird and Sophia Roux, *Mechanics before Scientific Revolution*, Netherlands: Springer Publications, 2008, p. 76.
13 Laird and Roux, p. 91.
14 Laird and Roux, p. 63.
15 William Heytesbury, *On Maxima and Minima: Chapter 5 of Rules for Solving Sophismata, with an Anonymous Fourteenth-century Discussion*, John Longeway (trans.), Dordrecht: D. Reidel Publishing Company, 1984, p. 64.
16 C.B. Schmitt, Q. Skinner, E. Kessler, and J. Kraye, *The Cambridge History of Renaissance Philosophy*, UK: Cambridge University Press, 2003, p. 221.
17 It is called "The Merton Rule of Uniform Acceleration." Edward Grant, *A Sourcebook in Medieval Science*, USA: Harvard University Press, 1974, p. 252.

18 Ernst Moody, "Galileo and Avempace: The Dynamics of the Leaning Tower Experiment", *Journal of History of Ideas*, Vol. 12, No. 3, June 1951, p. 375–422. The author shows that the line going to Galileo started with Bradwerdine who also depended on Ibn Bājja.

19 Laird and Roux, p. 93.

20 Oresme, *Tractatus of Nicole Oresme*, Marshall Clagget (trans.), Wisconsin: University of Wisconsin Publications, 1968, p. 164–7.

21 William Wallace, *The Modelling of Nature: Philosophy of Science and Philosophy of Nature in Synthesis*, USA: The Catholic University of America Press, 1996, p. 391.

22 Menso Folkerts and Richard Lorch, "Arabic Sources of Jordanus de Nemore", *Foundation for Science, Technology and Civilisation*, UK: FSTC Lmtd., 2007, p. 5–15.

23 Marshall Clagget, *Science of Mechanics in Middle Ages*, Wisconsin: Wisconsin University Press, 1952, p. 523.

24 A.C. Crombie, *Technics and Science in Middle Ages. History of Science*, New York: The Bobbs-Merrill Reprint Series, 1967, p. 213.

25 James Evans, *The History and Practice of Ancient Astronomy*, Oxford: Oxford University Press, 1998, p. 102.

26 Fred Kleiner, *Gardner's Art through the Ages: The Western Perspective*, Boston: Cengage Learning, 2010, p. 359.

27 Curt Sachs, *Our Musical Heritage: A Short History of Music*, USA: Greenwood Press, 1978, p. 117.

28 Philip Ball, *Elegant Solutions: Ten Beautiful Experiments in Chemistry*, UK: Royal Society of Chemistry, 2005, p. 13.

29 Moody, "Galileo and Avempace: The Dynamics of the Leaning Tower Experiment", p. 375–422.

30 A.C. Crombie, *Science, Art and Nature in Medieval and Modern Thought*, London: Hambledon Press, 1996, p. 97–8.

31 One of the most important studies on "functional-model" in the Islamicate Civilization was al-Jazari's work, shortly titled *Kitāb al-Hiyal*, that explained mechanical settings including many quantitative algorithms. This kind of modeling examined the mechanisms scaled to make a certain function completely. For examples see Durmuş Çalışkan, *Cezeri'nin Kitab'ül Hiyel'ine Mekanik Bir Şerh*, http://tekhnelogos.com/el-cezeri (18 March 2013). In fact, according to Kant, they learned nothing from the Eastern world but *mechanical arts, astronomy* and *numbers*. See John Zammito, *Genesis of the Kant's Critique of Judgement*, USA: University of Chicago Press, 1992, p. 39.

32 Lynn White, *Medieval Religion and Technology*, CA: University of California Press, 1978, p. 80.

33 *Regulator* was a very important mechanism in mechanical systems and it had abundant versions in Islamic history of mechanics.

34 A similar modeling was done at the same time by Kemal al-Din Farisī who was not familiar with Theodoric. Both of them made use of Ibn Haytam's drawings. See. Hüseyin Gazi Topdemir, "Kemaleddin Farisi'nin Gökkuşağı Açık- laması", *A.Ü. Dil Tarih Coğrafya Fakültesi Dergisi*, Vol. 33, No. 1–2, 1990, p. 472–92.

35 One of the early examples of crank and connecting rod mechanism was al-Jazari's water pump. Although this kind of examples were seen

previously, functional models fitting to working principle began to appear after al-Jazari in a structure with hitherto unseen complexity.

36 The works of some important philosophers such as Cicero (MÖ 106-43), Diogenes (MÖ 404-323), Lucretius (MÖ 99-55) can be counted as examples for this kind of translations.

37 Vitruvius's work was first printed in 1486 and in the introduction of the Chapter IX, the subtitle *"Platonis inventum agra metiendo"* was given. See. Vitruvius, *Architecture*, Liber IX, Caput 1, Giovanni Poleni (trans.), 1486, p. 7. With the phrase "Plato invented measurement" he paid due importance both to the concept of measurement and to Plato and linked them together. Similarly, just like Plato's academy, all academies appearing in whole Europe after the fifteenth century required the knowledge of geometry and arithmetic for enrolment.

38 J. Monfassani, *George of Trebizent, A Biography and Study of His Rhetoric and Logic*, Leiden: Trustees of Colombia University, 1976, p. 201.

39 Elspeth Whitney, *Medieval Science and Technology. Greenwood Guides to Historic Events of the Medieval World*, USA: Greenwood Publishing Group, 2004, p. 169.

40 Crombie, *Science Art and Nature in Medieval and Modern Thought*, p. 325.

41 John Hendrix, *Platonic Architectonics*, New York: Peter Lang Publishing, 2004, p. 169.

42 A.C. Crombie, *Science, Optics and Music in Medieval and Early Modern Thought*, London: Hambledon Press, 1990, p. 164.

43 Among the most important examples are Archimedes's translations (1544), Aristotle's *Mechanica* (1525, 1552) and *De audibilibus* (1562), Heron's *Pneumatica* (1575), Apollonius's *Conic Sections* (1537), Diaphontus's works on mathematics and algebra (1575), Ptolemy's *Geographia or Cosmographia* (1477) and *Harmonicorum* (1557), Theophrastus's *De sensibus* (1516) and *De historia plantarum* (1644) and Pappus's *Mathematika Collectiones* (1588).

44 Perspective originated from the need to show the depth while seeking factuality. Ottoman miniatures had some factuality but this depth could only appear by interpretation. While not contrary to the language of representation, the modern perspective could be perceived by a nonexpert, so it spread thanks to its operability.

45 W. Ashworth, "Natural History and Emblematic World View", in *Reappraisals of the Scientific Revolution*, David Lindberg and Robert Westman (ed.), New York: Cambridge University Press, 1990, p. 303.

46 The work written by Ibn Mu'min was translated by Theodor of Mopsuestia into Latin with all detailed under the title *De Scientia Venandi per Aves*.

47 Andrew Feld, *Raptor*, London: University of Chicago Press, 2012, p. 23. Frederick's own work *De arte venandi cum avibus* was published in 1241

48 C.H. Haskins,"The"De Arte Venandicum Avibus" of the Emperor Frederick II", *The English Historical Review*, Vol. 36, No. 143, July 1921, p. 334–55.

49 Frederick II of Hohenstauffen, *The Art of Falconry*, Marjorie Fyfe and Casey Albert Wood (trans.), Stanford California: Stanford University Press, 1943, p. Xlix.

50 Samuel Edgerton, *The Mirror, the Window and the Telescope*, Cornell University Press, USA, 2009, p. 39–40.

51 Giorgio Vasari, *The Lives of the Artists*, Julia and Peter Bondenella (trans.), Oxford: Oxford University Press, 1991, p. 114.
52 Elisabeth Gilmore Holt, *A Documentary History of Art: The Middle Ages and Renaissance*, Vol. 1, Princeton: Princeton University Press, 1981, p. 157.
53 Peter Burke, *Culture and Society in Renaissance Italy 1420–1540*, London: Batsford, 1972, p. 126.
54 Holt, p. 158.
55 John White, *The Birth and Rebirth of Pictorial Space*, 3rd edition, Boston: Faber and Faber, 1987, p. 134.
56 Nicholas Temple, *Disclosing Horizons: Architecture, Perspective and Redemptive Space*, New York: Routledge, 2007, p. 98.
57 Pamela Long, *Openness, Secrecy, Authorship*, London: Johns Hopkins University Press, 2001, p. 130.
58 Mark Smith, *Alhazen's Theory of Visual Perception*, Philadelphia: Transactions of the American Philosophical Society, 2001, p. 393.
59 Mark Smith, p. 480.
60 Mark Smith, p. 484.
61 Leon Alberti, *On Painting (De Pictura)*, John Spencer (trans.), UK: Yale University Press, 1956, p. 90.
62 Alberti, p. 112.
63 Crombie, *Science, Optics and Music in Medieval and Early Modern Thought*, p. 162. *Virtue*, originally comes from Greek *arete* whose meaning was used by the Stoics as the skills as combination of knowledge and wisdom.
64 Leon Battista Alberti, *On the Art of Building in Ten Books*, Joseph Rykwert, Neil Leach and Robert Tavernor (trans.), Cambridge: MIT Press, 1988, p. vi.8–x.11.
65 Angelo Mazzocco, *Linguistic Theory in Dante and Humanists*, Leiden: E.J. Brill Publishing, 1993, p. 104. The history of Islamic thought can also be given as an example to scientific and cosmological developments in vernacular languages. Since no problem existed in the whole system, Arabic concepts were kept in translations of the texts on logic and philosophy from Arabic to Turkish. The divergence between the relations of Latin-national languages and that of Turkish-Arabic had significant effect on this.
66 Kim Williams, Lionel March and Stephen R. Wassell (ed.), *Mathematical Works of L.B. Aberti*, Basel: Springer, 2010, p. 194–6.
67 Jeanne Pfeifer, "Projections Embodied in Technical Drawings: Dürer and His Followers", in *Picturing Machines 1400–1700*, Wolfgang Lefevre (ed.), USA: The MIT Press, 2004, p. 245–6. Dürer showed in detail the techniques he used in his studies in his work titled *Underweysung der Messung (The Painter's Manual)* (1525).
68 Factual description as a sociocultural distinction is the most important characteristic of the tradition of Islamic thought. The thing transmitted to the Western world was actually this characteristic.
69 *Agricola's De Re Metallica* had translations in all European languages and publications because of the fact that it was a book on mining as a basic need. Its publication continued even in the nineteenth century.
70 Ian Inkster, "Potentially Global: 'Useful and Reliable Knowledge' and Material Progress in Europe, 1474–1914", *The International History*

Review, Vol. 28, No. 2, July 2006, p. 246. Ayrıca bk. Francis C. Moon, *The Machines of Leonardo Da Vinci and Franz Reuleaux: Kinematics of Machines from the Renaissance to the 20th Century*, Dordrecht: Springer, 2007. Similarly, Da Vinci's precise drawings of human head, Vesalius's detailed drawing of human skull and heart tissues (*De humani corporis fabrica*, 1543); Felix Platter's (1567) drawings of popped eyes, Joseph Guichard Duverney's (1683) works on ear hearing mechanism had advanced techniques of drawing. See Andrea Carlino, *Books of the Body: Anatomical Ritual and Renaissance Learning*, Chicago: University of Chicago Press, 1999, p. 44–49.

71 Samuel Edgerton, *The Heritage of Gitto's Geometry: Art and Science on the Eve of the Scientific Revolution*, Ithaca and London: Cornell University Press, 1991, p. 258–63.

72 Vitruvius, *Ten Books on Architecture*, p. 5.

73 Vitruvius, p. 9–12.

74 Vitruvius, p. 283.

75 Vitruvius, p. 283.

76 Vitruvius, p. 284.

77 Gary Hatfield, "Metaphysics and the New Science", *Reappraisals of Scientific Revolution*, David Lindberg and Robert Westman (ed.), New York: Cambridge University Press, 1990, p. 109.

78 R.W. Serjeantson, "Proof and Persuasion", *Cambridge History of Science: Early Modern Science*, Vol. III, Kathrine Park and Laroine Daston (ed.), USA: Cambridge University Press, 2006, p. 151.

79 James Lattis, *Between Copernicus and Galileo: Christoph Clavius and the Collapse of Ptolemic Cosmology*, Chicago: The University of Chicago Press, 1984, p. 31.

80 Lattis, p. 31.

81 Lattis, p. 34–5.

82 Lattis, p. 32.

83 Mark Schiefsky, "Art and Nature in Ancient Mechanics," in *The Artificial and the Natural*, B. Vincent and W. Newman (ed.), USA: Massachusetts Institute of Technology, 2007, p. 106.

84 Jessica Wolfe, *Humanism, Machinery and Renaissance Literature*, Cambridge: Cambridge University Press, 2004, p. 43–4.

85 Crombie, *Science, Optics and Music in Medieval and Early Modern Thought*, p. 410.

86 Wolfe, p. 81.

87 Arianna Borrelli, "The Weatherglass and Its Observers in the Early Seventeenth Century", in *The Philosophies of Technology: Francis Bacon and His Contemporaries*, Claus Zittel, Gisela Engel, Nicole C. Karafyllis and Romano Nanni (ed.), Leiden: Brill, 2008, p. 68.

88 J. A. Chaldecott, "Bartolomeo Telioux and the Early History of the Thermometer", *Bulletin of the British Society for the History of Science*, Vol. 1, 1952, sp. 215–16.

89 Borrelli, p. 100.

90 Borrelli, p. 101.

91 Borrelli, p. 103.

92 Robert Campbell, *The London Tradesman*, London: T. Gardner, 1747, p. 245. When Galileo started work for the first time he was head of a

technician group making tools for the professors teaching mathematics as part of *quadrivium*. His close relation with tools began by this occasion. "Mathematical Instrument Maker" grew to be the most respectable occupation in Europe within a century.

93 The relation between the light's angle of refraction and the environment had already been known. Ibn Sahl (984) and Ibn Haytham, who developed the works of optics to a more advanced level, showed geometrically these angles of refraction, but which parameters that the relation was based was explained by Snell (1602) as "the law of refraction."

94 Although the ideas of "epistemological revolution" or "turning point" have been predominant in the history of science, some approaches explaining the process with as a continuity with the concepts of *intensification/dilution*. See. İshak Arslan, *Çağdaş Doğa Düşüncesi*, İstanbul: Küre Yayınları, 2012, p. 26.

95 Peter Harrison, "The Book of Nature", *The Book of Nature in Early Modern and Modern History*, Klaas van Berkel (ed.), Leuven: Peeters Publishers, 2006, p. 24.

96 John Henry, *Scientific Revolution*, Hong Kong: Palgrave Macmillan, 2002, p. 22. In fact, Galileo's position before being *Natural Philosopher* at the court palace was *mathematicus*. Those making mathematical tools as part of *artes serviles* were considered to be of lower classes.

97 *Dynamics* as a subsection of the classical mechanics examines the changes of movements of objects under various forces whereas *Kinematics* examines the object's speed, acceleration and the relation between them by disregarding the factors causing movement.

98 William Wallace, "The Influence of Aristotle on Galileo's Logic and Its Use in His Science", *The Impact of Aristotelianism on Modern Philosophy*, Ricardo Pozzo (ed.), USA: Catholic University Press of America, 2004, p. 70.

99 In the discussion of Aristotle's *Posterior Analytics*, in order to have a true statement, the premises must be true, they should be primary and not need any representation but be self-proven.

100 William Wallace, *Galileo and His Sources*, Princeton: Princeton University Press, 1984, p. 20.

101 Wallace, *Galileo and His Sources*, p. 114.

102 Crombie, *Science, Art and Nature in Medieval and Modern Thought*, p. 201.

103 Crombie, *Science, Art and Nature in Medieval and Modern Thought*, p. 204.

104 Eamon, "From the Secrets of Nature to Public Knowledge", *Minerva*, p. 324.

105 A.C. Crombie, *Science, Optics and Music in Medieval and Early Modern Europe*, London: Hambledon Press, 1990, p. 52.

106 Crombie, *Science, Optics and Music in Medieval and Early Modern Europe*, p. 53.

107 The fact that modern knowledge is *publicly approved* meant to test it for learning and teaching and to present it in mathematical language. *Ta mathemata, manthanein* comes from the verb to learn and *mathesis* from to teach. For Heidegger's discussion on this see. Heidegger, *On Science*, ed. Trish Glazebrook (Albany: State University of New York Press, 2012). It is worth noting that Heidegger perceived *mathemata* as "taking." If the taking does not occur, so does not the learning, because human being learns only what he/she owns. For someone deprived or suffering, the only solution is to take. To teach is nothing but allowing others learn.

Departing from this classification of Heidegger, the occupation of *privy secretaries* can be described as "limited and controlled allowing."

108 John Henry, *Scientific Revolution*, Hong Kong: Palgrave Macmillan, 2002, p. 27. It is worth noting that Descartes became familiar with mechanical arts after his discussion with Isaac Beekman and from that time onward he focused on application mathematics on other fields. Descartes carried this interest to the field of military engineering in the army.

109 İhsan Fazlıoğlu, "İki Ucu Müphem Bir Köprü: 'Bilim' ile 'Tarih' ya da 'Bilim Tarihi'", *Türkiye Araştırmaları Literatür Dergisi, Türk Bilim Tarihi*, Vol. II, No. 2, 2004, p. 9–27.

110 Patricia Fara, *Science: A Four Thousand Year History*, Cambridge: Cambridge University Press, 1998, p. 103.

111 John Dee, *Preface to Elements of Geometry*, London, 1570, p. 1. http://www.gutenberg.org/files/22062/22062-h/main.html (25 Ocak 2013).

112 Dee, p. 1.

113 Dee, p. 1.

114 Dee, p. 2.

115 Dee, p. 20–1.

116 Dee, p. 1. Dee's definition of *Archemastrie* is the following:

> This Arte, teacheth to bryng to actuall experience sensible, all worthy conclusions by all the Artes Mathematicall purposed, & by true Naturall Philosophie concluded: & both addeth to them a farder scope, in the termes of the same Artes, & also by hys propre Method, and in peculier termes, procedeth, with helpe of the foresayd Artes, to the performance of complet Experiêces, which of no particular Art, are hable (Formally) to be challenged.

> In addition, for the appraisal on Archytas's flying artificial goose as magic see John Wilkins, *Mathematical Magick or Wonders that may be performed by mechanical geometrie*, London: Edward Gellibrand, 1648, p. 191.

117 Giambattista Vico, *On the Most Ancient Wisdom of Italians*, Lucia M. Palmer (trans.), London: Cornell University Press, 1988, p. 45.

4 The Space of the New Knowledge

4.1 The Possibility of Storable and Publishable Knowledge: Kunstkammer

Kunstkammer and *Wunderkammer* rose as centers of storing and publishing knowledge starting in the seventeenth century. In German, *Kunst* means "art," *Kammer* means "chamber" or "boxroom" and so *Wunderkammer* means "chamber of curiosity." In these chambers, many works including stuffed inanimate animals, various fossils, plants, precious gems, remnants of mythological animals such as unicorn or dragon mostly, various mechanical clocks and numerous automat settings were displayed. In fact, certain sciences such as botanic, zoology and mineralogy began to develop by means of museums, they were classified, in Foucault's words, as "natural history."[1] *Kunstkammers* represented the different character of the new science for they included the idea of being informed of far distant places by politically powerful people and the sense of controlling.

Kunstkammers should be examined as a preparatory stage leading to laboratories for the systematic production of experimental knowledge. Thanks to geographic discoveries especially, from the fifteenth century onward, various natural and artificial objects from farthest lands of the world in the East and the West began to be collected and displayed in a space or cabin. Collecting formed physical ownership on knowledge through the objects collected. The fact that especially higher echelons of the society, like merchants or nobles, collected or had others collect these kinds of objects and established *Kunstkammers* and displayed them there gave them some kind of superiority or allowed them to keep their social status.[2]

Following the works of Plinius (23–79) on natural history, Ulisse Aldrovandi (1522–1605) and Athanassius Kircher (1602–1680) were the first persons who worked on the subject. Since information that did/

DOI: 10.4324/9781003275756-5

could not have space in classical books were transmitted through *Kunstkammer* objects, they had meanings beyond the objects themselves. With the idea and desire to own natural objects as well as exotic artifacts, *Kunstkammer* played a critical role in having shared space for natural and artificial objects and filled the intersecting space between *tekhne* (art, technics) and *natura*. Although isolating natural objects from their own context and identifying them with artificial objects was the indicator of a new understanding of ownership, it was based on a more essential foundation. The irrational perception coming from ontological insecurity and the settlement with the practice rendered the outer setting as an element of competition through a new perception of sovereignty by rationalizing it.[3] The power accumulation of the local forces against the authority depended on their raising the competition to the universal level and defining the universal or taking it with difference or pushing it to the universal level (Figure 4.1).

The facts that *Kunstkammer* referred to art through *Kunst* and that it worked with an understanding of art that was made not only of artificial objects but also of rare objects of nature showed how the putative

Figure 4.1 An example of *Kunstkammer,* engraving from Ferrante Imperato, *Dell'Historia Naturale,* Naples: 1599, retrieved from wikimedia.org.

ontological demarcation between nature and art was transitive. Besides, artistic applications on interesting geometries of some stones were also indicators of the two-way transitivity. These collections were theater displaying nature and art together as well as their old and new relations according to the view and order of their physical looks.[4] Collecting natural objects was effective in small political units because it was less costly and more remarkable than the artifacts that received intensive and systematic patronization during the Renaissance.

From this perspective, although some difference was somewhat possible between the sixteenth-century humanist collections and the seventeenth-century mercantilist collections, natural and artificial objects seemed to achieve a common look by the end of the century.[5] It was possible to take the collections of artificial objects, an alternative to the collections formed by the court's patronization, as an outcome of the mercantilism. The collections launched by the court as source of legitimacy and power accumulation were reestablished and reproduced by the bourgeoisie at cheaper cost and with more colors. As *Kunstkammers* in early modern period catered to those who were interested in them, it was aimed that they be a *micro cosmos* or *universal theater*.[6]

According to Lorraine Daston, *Wunderkammers* had an effect in removing the ancient distinction or conflict between nature and arts.[7] As rare objects of nature were discovered, their artistic aspect brought up the assumption that they became prominent by one of their particular aspects or privilege; therefore, nature made things by an artistic language. This can be interpreted as that human being saw nature as rival to himself. As an example to the efforts of controlling nature by arts, one can mention Palissy's (1510–1590) desire to make ceramic and metal versions of many plants and animals modeled by their animate versions. This meant the complete imitation of nature "without using the forms or methods of any arts or handworks of any artists" similar to how nature would turn fossils into rock. It is interesting that Palissy argued in the introduction of his work written in dialogue form that practical knowledge rather than theoretical knowledge of Latin philosophers was required to talk about nature.[8] Again similarly, Aldrovandi was collecting remnants of rock of cats, birds, dogs and human beings that "nature petrified."[9] It was clear that these objects were formed with an art by nature.

Similarly, "Solomon's House" Bacon mentioned in *New Atlantis* would include a cabin containing fruit trees with distinctive artistic characteristics in addition to automats, mathematical tools and normal nature of numerous gardens and structures.[10] Bacon rejected the

Aristotelian distinction between natural and artistic qualities and drew attention to the transitivity between the two and argued that nature deviated from its own way in both *mechanical arts* (*artes mechanicae*) and artistic structures on natural objects.[11] This idea of "deviation" brought a new understanding of order. Magnificent structures in nature were the artistic products of nature and should be welcomed for they guide human arts and new discoveries.[12]

Displaying these objects were important for showing the modern human relationship with nature. A relation with these objects through natural ways would of course have potential to break the process of this obtaining and displaying. The productions in both fields had mutually equal wonders and objects were displayed by human beings as imprisoned. Treating nature as captive also showed the human insecure relations with his own nature. Freed from being a threat by modern methods, nature was moved to a secure field with some control and observation. This ontological insecurity played a critical role with the chemical bond between *tekhne, natura* and *scientia* formed by modern science. The fact that *Kunstkammers* were encouraged by political authorities was because of their potential influence on the addressed audience. The political authority obtained legitimacy and provided security to visitors through this privileged exhibition for which it collected and possessed objects.

Every object in *Kunstkammer* as a microcosmos of the whole universe was detached from their natural contexts and a play was put on stage according to the role human beings assigned.[13] The one who possessed and controlled this microcosmos would also hold the monopoly of "understanding", "explaining" and "directing" this macrocosmos. Therefore, the fact that *Kunstkammer* provided a potential determining the individual's ontological position placed it to a critical position. The tradition of *Kunstkammer* gained prominence in respect to the owner's property relations to it rather than to the perspectives of visitors. The privileged knowledge and patent systems in that respect concur structurally with *Kunstkammer.*

4.2 Laboratory as Knowledge Workshop

Since a sharp distinction between "knowledge" and "thought" existed in the ancient period, public acceptance of the knowledge obtained through speculative philosophy could become possible only when they were formulated as geometric postulates. As the phenomenal world was chancing in respect to time and conditions, the produced knowledge needed adjustment in new conditions. The empiricists seeking certain knowledge however proposed a possibility that could appear in experimental conditions, which indicated a characteristic different from classical distinction of certain knowledge-idea. As every knowledge was always open to question could reach a kind of certainty within this ambiguous framework despite no connection required. This method, that is, the idea of "forming epistemological integrity through some reserves," was quite new. The classical understanding of nature, on the contrary, either rejected the possibility of this kind of knowledge or placed some physical distinctions to the status of "unchanging and certain knowledge."

The relations between modern science and experiment was quite critical for showing its departure from the Ancient understanding of knowledge mentioned above and at the same time from its understanding of nature. The new experiment should be seen in some sense as the incubation system of the new knowledge, and so should the laboratory as the possibility and beginning of its systematic production. In fact, the new experiment not only revealed the forces of nature but also allowed them being used for certain direction and purpose. The processes in the classical alchemy room were never seen as a meaningful relation for an observant. Besides, as the background of a new mixture obtained by alchemy was not known, it could never provide clear knowledge on nature like a telescope or microscope.

Even if it looked similar to the classical place where alchemist worked to produce, the laboratory had solid contributions for the modern new science's reaching its *sui generis* character. Although *laboratorium* was a Latin word for its expression, it had no reflection in the classical world. Its closest use to the modern sense appeared from the sixteenth century onward. Ophir and Shapin argue that the modern Western science was produced in heterotypic spaces. According to this, the laboratory was a heterotypic space in which limited number of different social actors were working together, so the things not seen otherwise made themselves visible here and the scientific processes processed on certain phenomena revealed (or were forced to reveal) themselves here. However, this visibility was applicable only to those who had the authority to be present in these environments. The laboratory became a space allowing both social abstraction from the rest of the world and epistemological construction of the visibility.[14] It would not be wrong to say that the continuous inclination to a certain problematic behind the walls of a limited space represented the mentality fitting to spatial concentration of the period. These spaces not only provide a secure zone to the insiders but also reflect the effort of the established settings to isolate from the environmental circumstances as much as possible.

The fact that the space in which the new experimental program was produced was open to public interest and accessibility asserted that the thing researched was attained by the common wisdom. The laboratories of the many of the seventeenth-century philosopher scientists were established in a room on the ground floor of their houses. Robert Boyle generally kept this door open and put note on the door if he wished himself undisturbed.[15] Although the privileged knowledge system was operating for three centuries, the open-door policy of the laboratories provides us different implications. One can state that outputs of a philosopher of nature were the building blocks of a level of consciousness contributing to the problem of ontological security rather than to a commercial commodity.

According to Boyle, the correspondence of an experiment's output to the reality depended not solely on its performance but also on its results' acceptability by the target group. Just as a certain man's committing to murder could become sufficiently worth of research by the existence of two witnesses, the existence of more than one witness to experimental outputs would be another factor increasing their reliability.[16] It is worth noting that the need for the approval by a second man or others, for the interest of the one producing these outputs, as well as the outsiders' paying attention to the outputs show the social reflection of the experimental science.

As noted above, while the experimental knowledge was produced for the first time in its particular spaces and within its own structure, it needed to go through the mechanism of witnessing open to public scrutiny. Shapin argues that the experimental knowledge would become enduring thanks to this circulation.[17] Faraday's method could be an important example to how much theatrical displays contributed to understanding what scientists did. David Gooding mentioned in his work on the role of these displays in the perception of Faraday's "electromagnetism" and his effort to convince other scientists around himself. According to Gooding, Faraday hid the background work at the displays and allowed the audience to have direct contact with nature and reworked the magnetism in respect to the feedback coming from the audience.[18] Thanks to the need for public approval, the experiment or exhibition could rise on stage only if it had solid conceptual ground and applications based on experiences in reproducible formats.

Machina Boyleana, also known as air pump, in order to spread the prestige of Royal Society to the whole continent, evolved into a tool having visitors enjoy exciting sessions. Boyle's air pump, as the most favorite instrument of the institution, appeared as the first comprehensive setting-device after the whole alchemy tradition. According to Boyle, the possibility of obtaining knowledge through senses by using these instruments was much greater than obtaining it through senses only.[19] The fact that even such a setting attracted the interest of all social elements including women and children and had almost no practical use indeed showed us a deeper problem with the perception of reality. It is possible to take these settings as portable laboratory. In fact, the instruments especially in early period were the systems composed of multiple parts (Figures 4.2 and 4.3).

According to Peter Dear, the failure of the experimental knowledge to appear with a method that could contribute to the philosophy of nature in Ancient Greece was because of the distinction between the natural and artificial, that is, *physis* (of nature) and *tekhne*. The objects and processes that could be part of a setting used for experimental purposes were not considered a consistent method at that time because they would be removed from their natural structures and brought into a debatable state.[20] According to Aristotle understanding nature depended on its observation without any intervention directly or indirectly whereas according to Francis Bacon obtaining knowledge on nature could be possible only by direct intervention.[21]

Figure 4.2 Magdeburg experiment of vacuumed globe. O. von Guericke, experiments with vacuum, 1672, retrieved from wellcome collection.org.

Figure 4.3 The scene of air pump experiment drawn by Joseph Wright in 1768, National Gallery, London, retrieved from wikimedia.org.

According to Shapin, the thing encouraged, and even forced, Robert Boyle to keep the laboratory's door open was social demands and expectations.[22] Rendering the produced knowledge durable could only be possible by this kind of public relations. As Boyle was an English noble person, keeping doors always open to gentlemen was also related to the social order. Although the seventeenth-century books on manners often stated that being gentlemen could last only with hospitality, this opportunity was closed to lower classes in Boyle's case.[23]

Boyle's laboratory became a center of frequent attraction especially for foreign visitors. A Florentine experimentalist Lorenzo Magalotti stated in 1668 that he went to Oxford especially to visit it and received information equivalent to 10 hour class and John Evelyn asserted that Boyle spend the afternoons in the laboratory generally with a foreign visitor or a friend.[24]

As a natural result of these relations, the knowledge produced within the laboratory communicated somehow with the external world. By the eighteenth century, this kind of knowledge began to be used also for attracting social attention.[25] For example, James Graham's presentation of electric experiments on the stage as a sacred show of power

Figure 4.4 A satire of lecture on pneumatics at the Royal Institution, coloured etching by J. Gillray London: 1802, retrieved from wikimedia.org.

was important for showing the theatrical aspect of modern science.[26] As the events and processes that seemed normal in laboratories turned into a mysterious display in front of crowded audience, similar applications were carried out in sessions addressing to socially elite groups filled mostly by women at the auditoriums of various scientific institutions.[27] Coffee houses, entertainment houses, meeting houses, publicly open centers were leading spaces in which this new entertainment and show concentrated. A century later, the universities offering public lectures became the institutions that kept addressing to this interest within the discipline (Figure 4.4).

4.3 The New Knowledge's Artificial Nature: Mechanical Arts

To explain the production processes of the new knowledge and the sociopolitical reflection of this produced knowledge it would be sufficient to look at Robert Hooke and Francis Bacon. At this stage, Hooke can be taken as the person who produced experimental knowledge and did this especially through mechanical tools, whereas Bacon appeared to be the person who transformed perfectly the productions of the contemporary philosophers of nature into a part of sociopolitical system.

In respect to the knowledge-senses relations, Hooke points out that mechanical inventions can possibly improve our senses such as hearing, seeing and smelling. As an example, he argues that the sounds that cannot be heard through natural hearing organ can be perceived thanks to the tools developed in the future; and that with the help of a tube he has developed, he could transmit the sound to the receiver with little effort.[28] Hooke affirms that the arts of life that have been too long imprisoned in the shops of mechanics have been freed, thanks to the munificence of Sir John Butler who endowed a lecture for the promotion of the mechanical arts (*mechanick arts*) to be directed by the Royal Society. He argues that Butler has taught (and even obliged the merchants for this) how he has done the trade worthy of London and taught the chief city of commerce in the world the right way how the commerce is to be improved.[29]

While examining small animate beings by microscope and defining the difference between nature's art and human art, Hooke pronounced that the one belonging to human was "crude and deformed."[30] The views of complex and homogenous tissues that Hooke obtained through microscope kept intact the hypothesis that everything could have a different essence. All of works based on both microscope and telescope aimed at going beyond the emblematic perceptions by applying the principle of factuality (Figure 4.5).

Hooke stated that they eliminated many mistakes and obsessions from the past thanks to the experiments closely witnessed by the members of the Royal Society. Since people begin to find how those effects of Bodies commonly attributed to occult qualities are "perform'd by the small Machines of Nature [...]; so as now they are no more puzzled about them, then the vulgar are to conceive, how Tapestry or flowred Stuffs are woven." These noble members do not reject pure optics experiments as long as these experiments serve to improve the applications of the manual arts. Once the capital is added to the acquisitions within the Royal Society by the next step, they would satisfy the sustainability criterion. As a matter of fact, Hooke states,

> that several merchants, men who act in earnest, have adventur'd considerable sums of money, to put in practice what some of our members have contrived, and have continued stedfast in their good opinions of such indeavours, when not one of a hundred of the vulgar have believed their undertakings feasible [...] that their attempts will bring philosophy from words to action.[31]

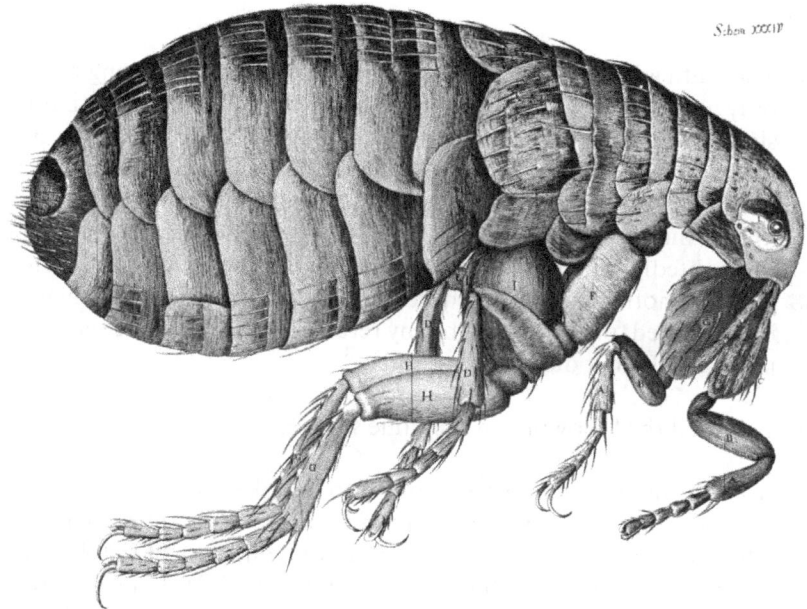

Figure 4.5 An example of anatomic drawing from Hooke's book *Micrographia*, Schem. XXXIV, 1665, retrieved from wikimedia.org.

Hooke states in the introduction of his *Micrographia* that Dr. Wren created in the king's closet a great collection of rarities (*Kunstkammer*) composed of the most precious works of the time. According to him, Dr. Wren was a unique person, after Archimedes, who held a philosophical mind and mechanical hand at the same time.[32]

What Hooke tried to describe at that time was to explain mechanically especially two useless concepts of the Ancient Peripatetics, that is, form and matter as they were accustomed to understand, the composition of objects, the structure of parts, the tissue of matter, the instrument of internal movements and all possibilities of their views. Therefore, he said,

> we may perhaps be inabled to discern all the secret workings of Nature, almost in the same manner as we do those that are the productions of Art, and are manag'd by Wheels, and Engines, and Spring devised by humane Wit.

Then he continued,[33]

> The truth is, the Science of Nature has been already too long made
> only a work, of the Brain and the Fancy: It know high time that it
> should return to the plainness and foundness of Observations on
> material and obvious things. It is said of great Empires, That the
> best way to preferve them from decay, is to bring them back to the
> first Principles, and Arts, on which they did begin. The same is
> undoubtedly true in philosophy, that by wandring far away into
> invisible notions as almost quite deftroy'd itself, and it can never
> be recovered, or continued, but by returning into the same sensible
> paths in which it did at first proceed.[34]

Hooke stated the following on the same subject:

> In both these the middle wayes are to be taken, nothing is to be
> omitted, and yet every thing to pass a mature deliberation: No In-
> telligence from Men of all Professions, and quarters of the World,
> to be flighted, and yet all to be so several examin'd, that there re-
> main no room for doubt or insatiability; much rigour in admit-
> ting, much strictness in comparing and above all, much slowness
> in debating, and shyness in determining is to be practiced. The
> Understanding is to order all the inferior services of the lower Fac-
> ulties; but yet it is to do this only as a lawful Master, and not as a
> Tyrant. It must not incroach upon their Offices, nor take upon it
> self the employments which belong to either of them. It must watch
> the irregularities of the senses, but it must not go before them, or
> prevent their information. It must examine, range, and dispose of
> the bank which is laid up in the Memory; but it must be sure to
> make distinction between the sober and well collected heap, and
> the extravagant Ideas, and mistaken Images, which there it may
> sometimes light upon. So many are the links, upon which the true
> Philosophy depends, of which, if any one be loose or weak, the
> whole chain is in danger of being dissolv'd; it is to begin with the
> Hands and Eyes, and to proceed on through the Memory, to be
> continued by the Reason; nor is it to stop there, but to come about
> to the Hands and Eyes again, and so, by a continual passage round
> from one Faculty to another, it is to be maintained in life and
> strength as much as the body of man is by the circulation of the
> blood through the several parts of the body, the Arms, the Fat, the
> Lungs, the Heart and the Head.[35]

However, Descartes defined for some time before the lack of confidence on senses as,

> I will shut my eyes, stop my ears, and withdraw all my senses. I will eliminate from my thoughts all images of bodily things, or rather, since this is hardly possible, I will regard all such images as vacuous, false and worthless.[36]

Descartes' turning into such a sharp rejection that could eliminate the whole biological self was an indicator of not only an effort to escape from epistemological sophistry but also a departure from a completely *ontological insecurity*. This insecurity rendered experimental tools as guide to in some sense limited natural organs of senses in examining natural processes through the way going to the scientific revolution.

Francis Bacon rendered an important comparison between science and philosophy in order to show theory-practice relations.[37] While the acquisitions of the first inventor fell short, these ideas were developed in time and thereby tools and objects evolve to perfection. As for science and philosophy, the texts penned by first authors tended to degenerate in time. Therefore, certain arts such as armoring, navigation, printing press at the beginning were conducted so crudely, but they attained the most perfect forms. However, the ideas and thoughts of philosophers such as Aristotle, Plato, Hippocrates, Euclid and Archimedes were pure and perfect at the beginning but they in time degenerated and changed. Just as water cannot reach a point higher than its source, an idea derived from Aristotle cannot reach a more advanced stage than his.[38] Implying that practical life and use should actually be prioritized in respect to the existing philosophical speculation, Bacon managed to raise a serious objection by mechanical arts to the hitherto predominant Aristotelian nature-art distinction.[39]

Themo Judaei, a follower of Jean Buridan from medieval Paris school had works and definitions focusing on reproducing natural processes experimentally that could contribute to the distinction between natural and artificial. Just as rainbow could produce halos (*halo artificialis*) by reproduction through arts, Themo asked similarly whether metals could be reproduced through arts.[40] In explaining another meteorology problem, the author stated that rainbow, based on the Aristotelian nature-art relations,[41] could be reproduced through "the art helping nature" (*artem iuvantem naturam*).[42] Thus, it can be said that Themo was a continuator of a tradition that treated practical experiences and applications over mutual relations in the nature-art distinction and that would later reserve space to experiment and settings in the philosophy of nature.

Contrary to the widespread appraisal that Francis Bacon had tendency to take initiative and intervene over nature by a capacity coming from alchemy, the primary factors were mechanical arts (*artes mechanicae*) such as navigation, compass and gunpowder.[43] According to Bacon, there were three kinds of nature:

> Nature in its unrestrained state (natura in cursu), nature in an accidental condition, as in the production of monsters (natura errans), and nature as "constrained, moulded, translated, and made as it were new by art and the hand of man" (natura vexata). Indeed, it is natura vexata (nature vexed) that corresponds to Bacon's call for an interventionist experimental science, and this will form a central focus of his reformed natural history.[44]

> Again, one might contemplate the subtlety which belongs to the liberal arts, or that involved in the preparation of natural bodies by the mechanical arts, and wonder at such things as the discovery of celestial motions in astronomy, of harmonies in music, of the letters of the alphabet in grammar; or, again in mechanics, the discovery of the products of Bacchus and Ceres, that is, the making of wine and beer, loaves of bread, or the delicacies of the table and distillation, and so on. One might reflect and notice how many centuries it took for these things to be brought to the state of development which we now enjoy, since all of them, except distillation, were ancient, and (as we said of clocks) how little they owe to observations and axioms of nature, and how easily, by ready opportunities and casual observations, they have been discovered. And thus, I say, he will easily free himself from all wonder, and rather pity the human condition, that through so many centuries there has been such a lack, such a dearth of objects and discoveries. And these discoveries which we have just mentioned are older than philosophy and the arts of understanding, so that (if the truth be told) since such rational and dogmatic sciences came into being, the discovery of useful works has ceased.[45]

Again according to Bacon,

> When a man makes the appearance of a rainbow on a wall by the sprinkling of water, nature does the work for him, just as much as when the same effect is produced in the air by a dripping cloud: and on the other hand when gold is found pure in sands, nature

does the work for herself just as much as if it were refined by the furnace and human appliance.[46]

Bacon continued to tell that mechanical arts gathered information from a wide spectrum and timespan -including natural and the present artistic applications:

> For a mechanic, who is by no means anxious about the investigation of truth, does not direct his mind or stretch his hand to anything but what is useful for his task. But the hope of further progress in the sciences will be well founded only when natural history shall acquire and accumulate many experiments which in themselves are of no use, but which simply help towards the discovery of causes and axioms; these we have been accustomed to call illuminating experiments as distinct from profitable experiments.[47]

For example, a crab's turning red when cooked was an important information; under normal conditions the importance of this knowledge might not readily appear but this experience could provide us knowledge on redness, similarly a brick would turn rend when cooked. Therefore, there was no condition that our researches always support individual arts such as agriculture, clockwork, carpentry etc. Thus, all mechanical experiments should flow from all direction into our sea of philosophy, no matter they could be of illiberal or lowly classes.[48]

Besides, Bacon asserted that this sea was a long-run program rather than a short-run, among the mechanicians the best educated ones would be supported, the would keep working as long as their products pay the principal investment, they would not chase after golden apple like children in order to make art dominate nature, they would not desire timeless and early results but aimed at a complete triumph in the long run.[49]

According to Bacon, human beings fought for three things. The first was one's effort to procure sovereignty over one's own country, the second was one's effort to drive one's country forward in the competition for humanity, the third was to effort to have domination over the universe and nature, which could only be possible by science and arts; and no power, *star* or *empire* could become as effective as mechanical arts on this.[50]

Creating rainbow by way of arts or nature does not belie the fact that it is always a natural product. The issue here was revealing the fact that natural processes that were not seen or sensed always could be observed by way of arts rather than determining the essence of artifacts.

Bacon meant this also by saying that arts were "human being added to nature" (*ars sive additus rebus homo*).[51] He stated that the history of nature and arts full of human intervention were in three kinds: by mechanical arts, by the practical side of liberal arts or by experiment and observation at the lower level with no arts.[52]

The idea of human being added to nature would allow concurrence between human acts and nature's acts. Bacon made the following comparisons on this subject:

> For the operations of nature are performed with much smaller portions and more precise and discriminating arrangements than in the operations of fire as now applied. Man would truly be seen to increase his authority if by heat and artificial forces operations of nature could be copied in kind, perfected in power and varied in number; to which should be added that they could be speeded up. Rust takes a long time to work on iron, but the effect of sesquioxide appears instantly; similarly with verdigris and white lead. Crystal takes a long time to grow to perfection, but glass is blown in a moment. Rocks take years to form, but bricks are quickly baked, and so on. Therefore (to return to our point) all the different varieties of heat with their respective effects should be diligently and industriously gathered from every source and investigated: celestial heat through rays, direct, reflected, refracted and concentrated in burning-glasses; the heat of lightning, flame, coalfire; fire of different materials; open fire, closed fire, forced fire, raging fire; fire modified by different furnace materials; fire excited by blowing, fire simmering and unstirred; fire at different distances; fire making its way through various media; damp heats, like Mary's baths, dung, the external heat of animals, the internal heat of animals and hay stored in a close place; dry heats, like ashes, lime, warm sand; in fact, every kind of heat with their degrees.[53]

Similarly Bacon reserved considerably more space to mechanical arts in another book and stated that the most useful history was art history, and arts (*artes mechanicae*) dealt with the moving things and more importantly they removed the veil on the face of nature, thereby natural processes were now easily observed.[54]

Another example showing theory-practice agreement was Paulo Rossi's argument that Francis Bacon's consideration of George Agricola possessing mechanical arts more superior than theoreticians such as Galileo and Copernicus was based on his view that logic was a tool to organize nature.[55] For Bacon, *Historie Mechanical* was the most needed thing on the way to the philosophy of nature, because this field dealt with the works of artists completely free from speculative philosophy.[56]

Besides, according to Hugh of Saint Victor, mechanical arts should be examined within philosophy. Four sub-sciences of philosophy were theory, practice, mechanics and logic. Theory addressed to reason, practice to values, mechanics to needs and logic to speaking.[57] The first human being had had three eyes before the Fall. According to Hugh's argument, since human being lost two of them, he had to use mechanical arts and philosophy in order to obtain eye of reason and eye of contemplation.[58]

It would last for a long time to work in a complete agreement and mutual support between the thing Bacon described as *scientia operative* and mechanical arts. During the classical period a mechanical art could use new outputs only within its own discipline whereas *scientiva operative* was an indicator of a new condition which brought together all kinds of technical and practical applications and appeared as a shared ground. Although Bacon's concepts of "newman of science" and "maker's knowledge" did not completely correspond to mechanical arts and the knowledge produced by these arts, it heralded that new relations should be established between theory and practice.[59]

Notes

1 Jan Golinski, *Making Natural Knowledge: Constructivism and the History of Science*, Chicago: University of Chicago Press, 2005, p. 97.
2 Later Leibniz proposed the establishment of an academy of science to display interesting and wonderful air pump settings, wondersome objects of nature, rare minerals, plants and animals. See Gottfried Wilhelm Leibniz, "An Odd Thought Concerning a New Sort of Exhibition", *Leibniz: Selections*, Philip Wiener (trans.), New York: Charles Scribner's Sons, 1951, p. 585–94.
3 The counter set encircling the individual can be defined as "external-individual." A connection between the individual and the external individual could be mentally possible.
4 Loraine Daston and Kathrine Park, *Wonders and Order of Nature 1150–1750*, New York: Zone Books, 1998, p. 265.
5 Daston and Park, p. 271.
6 Samuel Quichelberg's demand was organizing museums as a "universal theater." Similarly, French Doctor Pierre Borel hanged on his cabin the phrase "This is a microcosmos of rare things." See Daston and Park, p. 272.
7 Daston and Park, p. 261. In ancient world view, art imitated nature, in other words, it had a structure lower than nature, even it did not have an identity because it did not have a nature. No artists could go beyond imitation.

8 Louisa Dolza, "Technology and Culture During the Day of Bacon's France Stay", in *The Philosophies of Technology: Francis Bacon and Its Contemporaries*, Claus Zittel, Gisela Engel, Nicole C. Karafyllis and Romano Nanni (ed.), Leiden: Brill, 2008, p. 12.

9 Daston and Park, p. 286.

10 Daston and Park, p. 290.

11 Daston and Park, p. 291.

12 Sophia Weeks, "Mechanics in Bacons Great Instauration", in *The Philosophies of Technology: Francis Bacon and Its Contemporaries*, Claus Zittel, Gisela Engel, Nicole C. Karafyllis and Romano Nanni (ed.), Leiden: Brill, 2008, p. 159.

13 For analogy of theater stage, Aldrovandi's statement "Among Italian pharmacies the one that surprised me the most was Calzolari's theater of nature" was another example. See Daston and Park, *Wonders and Order of Nature*, p. 155.

14 See Adi Ophir and Steven Shapin, "The Place of Knowledge: A Methodological Survey", *Science in Context*, Vol. 4, No. 1, 1991, p. 13–14.

15 Steven Shapin, "The House of Experiment In Seventeenth Century England", *Isis*, Vol. 79, September 1998, p. 387.

16 Steven Shapin and Simon Schaffer, *Leviathan and Air Pump*, Princeton: Princeton University Press, 1985, p. 56.

17 Golinski, p. 95.

18 Golinski, p. 94. In addition, for Ludwig Fleck who proposed the concept of *Denkgemeinschaft* (collective thought) on the contribution of the audience and explained the formation of science through esoteric and exoteric connections, see Mark Erickson, *Science, Culture and Society: Understanding Science in 21st Century*, Cambridge: Polity Press, 2005.

19 Shapin and Schaffer, p. 18.

20 Peter Dear, *Discipline and Experience: The Mathematical Way in the Scientific Revolution*, Chicago: University of Chicago Press, 1995, p. 153–5.

21 Perez Ramos, *Francis Bacon's Idea of Science and the Maker's Knowledge Tradition*, Oxford: Clarendon Press, 1988, p. 50–5.

22 Shapin, "The House of Experiment in Seventeenth Century England", p. 387.

23 Shapin, p. 387.

24 Shapin, p. 386.

25 Shapin, by using Wittgenstein's concepts, argues the existence of a relation between "language game" or "life form" and experimental programs and activity patterns. Those who accepted that experimental method can produce knowledge would also accept Boyle's methods or vice versa. Epistemologically speaking, a subject or discussion depended on certain practical concerns and objectives. See Shapin and Schaffer, *Leviathan and Air Pump*, p. 15.

26 Golinski, p. 92.

27 Humphry Davy's various electrical and chemical exhibitions became one of the essentials of Royal Institution by the end of the eighteenth century. See *Golinski*, p. 93.

28 Robert Hooke, *Micrographia: Some Physological Descriptions of Minute Bodies*, London: Jo. Martyn, 1665, p. 11.

29 Hooke, p. 24.

30 Hooke, p. 2–8.

31 Hooke, p. 24.
32 Hooke, p. 25.
33 Hooke, "Preface," p. 8.
34 Hooke, "Preface," p. 9.
35 Hooke, "Preface," p. 10–11.
36 Rene Descartes, "Meditations on First Philosophy", *The Philosophical Writings of Descartes*, Vol. II, J. Cottingham, R. Stoothoff and D. Murdoch (trans.), Cambridge: Cambridge University Press, 1985, p. 24.
37 The fact that Kilwardby examined mechanical arts and some arts such as poetics and rhetorics as human arts showed the influence of translations from Arabic. The author who divided humanities outside of speculative philosophy as "verbal" and "practical" and classified the practical arts as "mechanical" and "ethical." According to him, practical arts were theoretical, theoretical arts were practical.
38 Francis Bacon, *Advancement of Learning*, London: Cassel and Company, 1893, p. 30.
39 Paulo Rossi, *Francis Bacon: From Magic to Science*, USA: Midway Print, 1978, p. 26.
40 Newman, *Promethean Ambitions*, p. 243. Themo argued that if metals were brought together with correct formula in their active or passive modes they could be obtained newer forms just like they did underground.
41 According to Aristotle, art (*tekhne*) completed nature.
42 Newman, *Promethean Ambitions*, p. 248.
43 Rossi, p. 26.
44 Newman, *Promethean Ambitions*, p. 258.
45 Francis Bacon, *The New Organon*, Lisa Jardine and Michael Silwerthorne (ed.), Cambridge: Cambridge University Press, 2000, p. 70.
46 Francis Bacon, *The Works of Francis Bacon*, Vol. 5, James Spedding, Robert Leslie Ellis and Douglas Denon Heath (ed.), London: Longman, 1857–1874, p. 507.
47 Bacon, *The New Organon*, p. 81.
48 Bacon, p. 228.
49 Bacon, p. 91.
50 Bacon, p. 100.
51 Francis Bacon, *The Works of Francis Bacon, Philosophical Works*, Vol. 3, James Spedding, Robert Leslie Ellis and Douglas Denon Heath, London: Houghton Mifflin and Co., 1887, No. 2, p. 731.
52 Bacon, *The New Organon*, p. 227.
53 Bacon, p. 214.
54 Weeks, p. 158.
55 Dolza, p. 7.
56 Jurgen Klein, "Francis Bacons's Scientia Operativa", *The Philosophies of Technology: Francis Bacon and Its Contemporaries* içinde, Claus Zittel, Gisela Engel, Nicole C. Karafyllis and Romano Nanni (ed.), Leiden: Brill, 2008, p. 22.
57 Hugh of Saint-Victor, "The Didascalicon of Hugh of St. Victor: A Medieval Guide to the Arts", Jerome Taylor (trans.), *Records of Civilization, Sources and Studies*, No. 64, New York: Columbia University Press, 1961, p. 9–22.
58 Victor, p. 4, 55, 56.
59 Klein, p. 25.

5 Mechanization

One of the problems frequently faced in the works of political history is the connection between individual contributions and historical progress. In this context, an example to people whose individual determinative role became prominent in the modern history of science and philosophy could be Francis Bacon, whose ideas were mentioned in previous chapters. The most important and known example of this was Royal Society of London, the most effective institution in the new directions Bacon initiated.[1] The members of the Royal Society were philosophers of nature who focused on the utilitarian science that Bacon called "new science," investors and members representing the royalty (Figure 5.1).[2]

In respect to the scientific revolution, the process from the mid-seventeenth century onward corresponds to a new period in which the feeling of deprivation partially disappeared and the significance of mechanical arts began to appear. During the timespan of approximately 150 years between Da Vinci and Galileo, creative and formative elements could not find the chance to institutionalize. The process starting from the mid-seventeenth century stood as the period in which the need for transforming these new dynamics of change into sustainable public structures. In this sense, since the "privileged knowledge system" was an essential element providing continuity for power accumulation, it would be transformed into reproducible and transferable structure through certain mechanization and would now be carried out as state project.

Besides that Bacon's contribution to scientific developments started with translation activities and conducted through the mechanical arts since the thirteenth century, the endeavors to carry out these new developments with a political agenda and by state apparatus and to generate their economic values were also considerably determinative factors. The systematization of the utilitarian science became possible

DOI: 10.4324/9781003275756-6

Figure 5.1 The Mace presented by Charles II which is placed on the table before the President at every meeting of the Royal Society, retrieved
from wellcomelibrary.org.

by the basic philosophical contributions it developed and the institutions formed in that respect.

According to Bacon, one of the mistakes one should avoid while gen-
erating an axiom based on realities is *Theatral Idol* (obsession). Various
systems and institutions—sometimes governments sometimes sophists
and pragmatists—captivate the human mind and become obstacles to
progress and innovations. Some men, known as philosophers, in order
to create a meaningless world, put these obstacles on stage, just like a
theatrical scene.[3] The experiments of the authorities who carried out
observations are problematic in respect to soundness and reliability.
The results and inferences derived from these experiments done behind closed doors are misleading and might lead someone into a vi-
cious cycle.[4]

It is striking enough to show Bacon's emphasis on mechanical science that he referred to the relationship between Daedalus[5] and the
mechanic. Daedalus, one of the craftsmen and artists, produced both
good and evil things with various tools and came out of his own lab-
yrinth with the wings he invented. As Bacon also states, Tacitus[6] too

said that once mathematicians and fortunetellers were banished from his country although they were hard to restrain through law.[7] When *artes serviles*, the arts usually carried out by lower classes, came into prominence due to competitions, these classes' esteem increased in the society.

Bacon demarcates the past and future clearly with his newly introduced method. Even though he occasionally refers to the "ancient," he argues that the modernity is such a new phenomenon which has not been tried before. According to Bacon, the new knowledge and the new method should be formed completely through the results of discoveries, not through ancient judgments.[8]

In Bacon's dreamed new world, the access to the publicly shared knowledge by all scientists throughout the world will be possible. As Eamon also states, however, *New Atlantis* is not totally a transparent society. Even though the "access to knowledge" seems to be free, interpreters of nature "will take a pledge on keeping secret" of the results of some experiments and observations, which, as they think, should be kept secret.[9]

For the first time in the history, there emerged a kind of organic setting in the modernity witnessing that multiple centers manage and lead the human mobility and receive the obtained results. In this context, the *Republic of Letters* emerged as a network making national borders transitional though respecting cultural and linguistic differences and working, in some sense, as an artistic institution.[10] The *Republic of Letters* was an intellectual institution making "effective close-distance interaction" possible and deriving its own legitimacy largely from the Royal Society. This communication network provided a data and strategy sharing not only among people but also between institutions.[11]

Bacon's works standing for the initiation of Royal Society were closely followed in Europe. The secretary of the institution, Oldenburg turned informal discoveries and flow of information among all scientists into a formal format. The fact that the periodical *Philosophical Transactions* brought reputation to the Royal Society stood for the evidence of free access to information. *Philosophical Transactions* and almost all similar periodicals shared the same platform where new philosophical understanding matured. This communication, which first started within the internal bureaucracy of the Royal Society, was not prevalent only among English scientists but it also contained whole Europe.[12] Obviously, the Royal Society's policy of recording and publishing was closely related to scientific discoveries and technologies of the time. As the spread of printing press was a critical instrument for realizing these goals, holding the monopoly of publication by certain individuals and institutions was also an important phenomenon. The fact that scientific discoveries and findings were subject to accreditation facilitated their attribution to true, real and appropriate scientists, and the centralization, reexamination and reproduction of knowledge.[13] On the other hand, the printing press contributed significantly to reproduction of practical experiences, which could hardly be transmitted through oral culture, in various regions of Europe at the same time.

Scientific practices and applications transformed into shared activities of certain groups of people thanks to mutual communication. These control mechanisms, which originated from a certain objective that can be summarized as "putting the accuracy of experimental results into warranty," were influential in this big transformation from very early on.[14]

As it is stated in the preface of Bacon's *Great Instauration*, the theoretical knowledge catches and manages nature on running, because it includes laws and principles concerning the nature. Thus, the mechanic is a science based on the principles of the nature.[15] This kind of knowledge produces itself or becomes meaningful in the technical application. This is the most important principle that Bacon and his followers inherited, which is "the coalescence of theory and practice or the development of practically applicable theories." The result of this coalescence will be "happiness" or "felicity" although this felicity is humanistic rather than divine.[16]

This happiness Bacon foresees will free human being from his/her deprivations and will give the knowledge (and thereby the control) of shared objects in the world. This control will also pave the way for the possibility of a certain (and probably infinite) use of the nature and objects, and thus the power acquired through knowledge will open a space of freedom to human being. This kind of power accumulation indicated an ontological state for the Western minds.

As Bacon states that the nature can be managed only by submission to its laws or the mastery over nature can be achieved only this way, he means that only the relations between objects can be changed and their substances, due to their unknowable aspect, cannot be replicated. He also argues that one can reproduce something existing in the nature if one sufficiently knows in detail its process of evolution, which leads us to the statement of "knowledge is power."[17]

According to Bacon who looks at the past through this perspective, "the man does not feel despair while looking at the past and seeing the little steps of progress from the antiquity to Renaissance; but feels pity for his previous deprivation from many facilities and inventions."[18] The knowledge should be produced "by adding new things to the nature of a given thing."[19] Thus, Bacon's perspective is expressed the best in this statement: the secrets of nature reveal themselves better through harassments applied by the arts than when they go on in their own way.[20]

The utilitarian model that has become systematic since Francis Bacon continued to interact with the "world of values" with its most comprehensive way in the twenty-first century. The capacity of producing this utilitarianism as constant values from its nature can become possibly only by the mechanization of its objectives. For example, according to Hans Jonas, the science in the classical theory used to deal with entities superior to human being, thereby "value" was perhaps the most basic tenet of the theory. Yet the modern scientist deals with inferior things, even treats stars and galaxies as objects. The answer for the question "how are they taken in respect to human being even if they are worthless?" could be "that which lacks intrinsic value of its own is lower than that by reference to which alone it may receive value, namely, man and human life, the only remaining source and referent of value."[21] Some sciences, like psychology, whose subject is directly human being seems to be candidate for having values because of their equal status in respect to human being. Yet since their interests are only natural man and unable to encompass the human being universally, they are objects of "scientific theory" and under the human being at some degrees; thus they are useful for human management and engineering.[22] Therefore, the science carries the benefit in itself, so it is technological.[23]

5.1 Scientific Tools as means of Applied Philosophy

The characteristic pragmatism of modern science brought the mechanisms giving correct results in subsequent tests into a central position. These mechanisms and instruments helped to verify the adopted philosophical system as well as to teach it. The new science instruments called "philosophical apparatus" usually imitated the natural processes corresponding to a utopic perception. [24] Even though the seventeenth-century scientific instruments, such as *camera obscura*, microscope, telescope, air pump, electrostatic discs, were not highly useful for daily life and practice, they had significant values in respect to groups of people, scientists or traveling merchants.

James Keir, a chemist, in his book *The First Part of a Dictionary of Chemistry* published in 1789, writes "The diffusion of a general knowledge and of a taste for science, over all classes of men, in every nation of Europe, or of European origin, seems to be the characteristic feature of the present age."[25] This approach is significant both because it witnesses that the science began to produce knowledge for all classes and because of the characteristics of new science.

Furthermore, the definition of "scientist" in a modern sense of the word was made by William Whewell (1794–1866) in 1830 for the first time. As the "science" was first among the phenomena intersecting all classes in the society, the natural philosopher who as working with an amateur perspective was now replaced by professional scientists.[26]

The experimental philosophy, which appeared in several institutions such as Royal Society, Accademia del Cimento throughout the seventeenth century began to be transferred into university system and acquired some independent chairs of science. Burchardus de Volder established a *Theatrum Physicum*, at Leiden University, where experimental exhibitions for air-pump were displayed.[27] John Keill (1621–1721) was able to initiate the teaching of mathematical and experimental philosophy of nature in England around the 1700s. Francis Hauksbee (1663–1713), who was tool assistant of William Whiston (1667–1752) (Lucas chair professor after Newton), published a work, *Physico-Mechanical Experiments* (in 1709), that was used in his classes made of experimental exhibitions and instruments.[28] Similarly, John Desaglier (1683–1744) joined the club of important persons for his lectures on experiments and his publications. It was him who translated many works from French and Dutch.[29]

The art of having classes with experimental exhibitions reached its peak with Jean-Antoine Nollet (1700–1770) who carried out *course de Physique* with 350 different instruments. Nollet met with a mathematician Polinere (1671–1734), at Academie des Sciences, whose classes on the philosophy of nature impressed him. After he met with Desaglier during his visits to England and Netherlands, he devoted himself completely to experimental physics courses. His lectures between 1743 and 1748 turned into a six-volume work, *Leçons de Physique Experimentale*.[30]

The fact that the science became popularized through instruments can be best exemplified by Benjamin Martin (1731–1821) who traveled cities and lectured with his experimental instruments, and by Adam Walker (1731–1821) who was an itinerant lecturer traveling with his "philosophical apparatus" including many mechanical powers, crane and pumps in Manchester and its vicinity, which was also advertised on a journal *York Courant* in 1772 (Figure 5.2).[31] These lectures were so effective that James Watt, the inventor of the steam engine, followed Willem's Gravesande (1688–1742) at the age of 15, and had the chance

Figure 5.2 Adam Walker and his family by George Romney, between 1796 and 1801, National Portrait Gallery, retrieved from wikimedia.org.

to obtain detailed information about the machines Newcomen atmospheric engines. Another example showing the importance of these lectures was that Martinus van Marum was charged by Netherland Haarlem state parliament to give lectures on philosophy and mathematics in 1776.[32]

The popularization of the experimental science in this way continued at universities and in various groups throughout the century and especially many people from noble and rich classes frequented them. Since experimental instruments and apparatuses became known by people through this way, people began to maintain them as individual property. This mobility raised London as a center selling experimental instruments to the whole world within a short time. As a result of this, considerable number of collections appeared in many countries such as America, Italy, Portugal and Denmark by the end of the century.[33] The shared channels opened by the imperialism also helped London to cater such a large geography.

The introduction of experimental instruments into curricula helped many artists to receive education. David Brewster states in his book *Lectures of James Ferguson*,

> We must attribute [to itinerant lecturers] the general diffusion of scientific knowledge among the practical mechanics of this country, which has, in great measure, banished those antiquated prejudices, and erroneous maxims of construction, that perpetually mislead the unlettered artist [artisan].[34]

As a matter of fact, these lectures aimed at educating individuals who had no competence on how to handle technical problems or how to use certain instruments. These mechanics would be the engineers with a university degree in the next century.

These lectures that continued throughout the century with no interruption usually covered the fields of astronomy, mechanics, magnetism, heat, pneumatics, optics and chemistry. Lectures on mechanics generally comprised of exhibitions of centrifugal forces, parabolic throwing and the set of force diagram made of levers, pulleys and cranes. In addition, explaining several mechanical subjects such as highly popular pulley and crane were explained through some apparatus; magnetic stone was followed with special interests because of its occult powers and its use in the compass was explained as well.[35]

If lectures happened to involve astronomy, it included models of the earth and telescopes, and planetariums modeling the solar system in advance levels of lectures. Hydraulics lectures introduced water fountains as a useful apparatus for gardening; pneumatics lectures presented the air pump as perhaps the most interesting apparatus of the century; lectures on heat introduced several instruments such as thermometer and hygrometer and Newcomen's machine as representing the steam power.[36]

The most important competitor for the two most popular instrument of the century, that is, "air pump" and "microscope," was the "electricity machine," developed by Martinus van Marum, based on the electrostatic electricity created by two revolving glass discs. A small version of this machine for domestic use and exhibition were highly spread (Figure 5.3).[37]

These scientific exhibitions and presentations popularized in the eighteenth century continued to be effective in mechanics for two centuries. William Whiston (1665–1752)'s drawings and designs (which Francis Hauksbee used in the introduction of his lectures in 1714 and included in his *Course*) appears again in a catalog prepared by

Figure 5.3 Traveling electricity experiment in the 18th century, historical engraving, 1880, retrieved from wikimedia.org.

J.J. Griffin in 1912. Similar designs began to be seen in school textbooks in 1960s. These eighteenth-century lectures were given for the middle age audience, and in the nineteenth century they were introduced into high school curricula and in the twentieth century into middle school curricula; and in the twenty-first century into the primary school curricula as part of scientific education.[38] Four centuries long process of the scientific revolution is interesting in respect to showing how far the modernity has been a determinant in the relationship between human being and knowledge. As the critical importance of the modern science paradigm in social structures was understood, especially the science's effect of ontologically security provision as an education policy began to be taken into consideration.

Summarizing the transformation in the eighteenth century, Nollet's work (*L'Art des Experiences, ou Avis aux Amateurs de la Physique, sur le Choix, la Construction, et L'Usage des Instruments*) explaining the art of experiment for amateurs was among the first important examples teaching how to make and use these instruments.[39] Various telescopes were ordered and used for research by various centers, especially for the use of transits of Venus, an important subject in the century. Thanks to reflecting mirror telescopes, this market developed significantly whereas microscope, on the contrary, continued to be used for domestic exhibition and entertainment and fail to develop for a long time.[40]

5.2 Continuity and Transmission of Knowledge: Official Curriculum

We need to show horizontal and vertical continuity of knowledge in order to talk about its continuity. Reproducible and transmittable information is readiest final form of knowledge that can be used for instruction. In this context, the systematization of particular aspects of modern science is related to the process of their introduction to curricula through universities. We observe the introduction of new science to curricula, which was an important phase of the mechanization, also as a process that shape and transform the curricula.

Until the eighteenth century, teaching subjects related to natural sciences had been done under the philosophy of nature. Many universities had a single philosophy professor teaching all branches of philosophy. Although important developments occurred in natural sciences, individual specialized professors had not been employed.[41] According to the established system throughout centuries, the universities usually had produced only personnel for the Church and legal system; thereby the curriculum had included only one-year physics.[42] Around the middle of the century, challenges to university curricula began to appear; demands for instructions in vernacular languages and for adaptation of mathematics and physics into contemporary needs increased. One of the known examples that addressed to these demands was philosopher Jean Le Rond d'Alembert (1717–1783).[43]

Even though, at the beginning, several reform attempts were made in various European countries toward the end of the century, it never went beyond the phase of trial. The demands were usually to establish Faculty of Arts, Mathematics and Physics. The first serious attempt appeared in France in 1795 with a special emphasis on natural sciences in schools named *ecole central* rather than *college*. However, due to the parents' complaints, these schools were transformed into *lycee*s prioritizing old curriculum and the universities were transformed into umbrella institutions named *Université imperial* in 1808. While a similar establishment appeared in Netherlands and Belgium in 1815, the foundation of independent faculties on science and the arts in the rest of the European countries could appear only in the middle of the century.[44]

Although the universities had a different venture after the beginning of eighteenth century, the formation of a new curriculum including sciences did not become prevalent until the middle of the nineteenth century. Since theology was still the most important and dominant field at the universities, philosophy and the arts were dealt with in the context of theology. Medical schools had almost no students in this period.[45]

Since Lutheranism was predominant in German universities, the classical connection between philosophy and theology faded from the seventeenth century onward, theologians began to marginalize both pietism and philosophical freedom.[46] As for Göttingen University, founded with a pluralistic understanding in the second quarter of the eighteenth century, the professors were able to choose textbooks and organize lectures by themselves as long as they did not contradict with religious and ethical values and the state. Even though all universities included Theology school, this would not define the dominant characteristics of a university. Universities would support to increase the number of publications by employing researchers from outside.[47]

Although new independent fields were not established in German universities by the end of the century, the philosophy of nature and some independent natural sciences began to obtain spaces within the established structures of the universities. Philosophy appear as an independent field in 1798 through the work (*Streit der Facultäten*) of Immanuel Kant'ın (1724–1804), a university professor at Königsberg. According to him, philosophy should be completely independent even although state should supervise other professional divisions because their products could be related to administration.[48]

As Kantianism became adopted generally, Philosophy gained an accepted position independent from Theology in all German universities. From the beginning of the nineteenth century onward, Theology schools also gained independent conditions. Newly emerging researches and seminars in natural sciences and mathematics now testified that universities began to dominate in these fields.[49]

The curricula of physics in European universities varied throughout the eighteenth century, as a result of spreading Newtonianism—largely with state promotions—in the whole continent, mathematical physics received a general acceptance by the end of the century. Descartes (1596–1650) and Gassendi (1592–1655), having some elements from the classical world view, were excluded from the curricula in many spaces, and a new system based on Newton principals and laws of gravity were established.[50] When theoretical physics in classical methods seemed to contradict with the new findings after 1670s, many professors began to seek help from outside. These experimental physics classes often turned into a kind of exhibitions on presenting experimental apparatuses and how to use them.[51]

Along with the popularization of experimental science, most of the departments of philosophy of nature were obliged to prepare laboratories composed of instruments. Thus, experimental physics gained a legitimate space in the curricula with a more institutional base. Otherwise, it at least survived in summer programs.[52]

Moreover, Baconian new scientific researches and new information and methods, particularly, in botany and zoology enabled to have a different point of view to the problems of life and reproduction. Thus, researches on human anatomy brought new perspectives and concluded that life could hardly be reduced to basic principles or to mechanical or mathematical models.[53] Yet, during the whole scientific revolution process, the strategy was based on the analogy between the nature and machine. Antoine-Claude Chaptal (1756–1832), a physician from the Napoleonic age, summarizes this situation:

> The laws of mechanics, hydraulics and chemical affinities act on all matter; but in the case of the animal economy, they are so completely subordinate to the laws of vitality that their effect is almost nil; and dependent on the intensity of that vitality, so living phenomena distance themselves further and further from the results calculated according to those [physical] laws.[54]

In addition, many occupational schools were established in order to fill the gap of technician personnel. By the end of the eighteenth century, occupational schools opened by governmental supports became widespread in whole Europe. The most important example among others was *Ecole Polytechnic*, established in 1794.[55]

Physics developed within two distinct disciplines in the eighteenth-century Europe. First, it was taught in the Faculty of Arts or Faculty of Philosophy as a mathematical phenomenon based on the Newtonian perspective and with a narrow curriculum, similar to Aristotle's *Physica* and *De Caelo*. Second, several physics subjects, such as mineralogy, chemistry, geology, zoology and anatomy, were taught in medical schools within the framework of a work, mistakenly attributed to Aristotle, because these subjects were not quantified yet and due to completely epistemological and pragmatic reasons.[56] While the experimental natural science that Bacon exalted appealed only to certain privileged groups in the previous century, this kind of science enabled ordinary people to join the system through universities and other institutions in the eighteenth century.

Although it was at the beginning just a field of occupation for normal university students who had attended mathematics classes, experimental philosophy turning science education into a theatrical entertainment and introducing new lecture techniques obtained a respected position. This kind of an extraordinary and esoteric approach to the nature coincided with entertainment and consumption patterns of rising social classes.[57]

At the end of this process, new science or experimental philosophy rose to a central position in the formation of European identity and consciousness. Even if Newtonian science could not completely replace the classical culture, it provided a minimum ground for the followers of new science. Thus, the booths for the philosophy of nature and antiques known as the most concrete element of new science were reserved in university libraries.[58]

Notes

1 Royal Society was Solomon House in Bacon's *New Atlantis*. See Michael Hunter, "A 'College' for the Royal Society: The Abortive Plan of 1667–1668," *Notes Rec. R. Soc. Lond* (London: 1 March, 1984), 160. Its title turned into Royal Society of London for Improving Natural Knowledge in 1663. It has been functioning in the same building for about four centuries.
2 R.H. Syfret, "The Origins of the Royal Society", *Notes Rec. R. Soc. Lond.*, April 1, 1948, p. 75–8.
3 Francis Bacon, *The New Organon*, Lisa Jardine and Michael Silwerthorne (ed.) (Cambridge: Cambridge University Press, 2000), 50.
4 Ibid, p. 18.
5 Francis Bacon, *Of the Wisdom of the Ancients* (New York: Kessinger Publishing, 1992), 245. He was the father of a Greek mythological hero, Icarus. For his son, he made wings from wax. Despite warnings, Icarus flew at height and the sun melted these wings, which caused Icarus fall into the sea and be drown.
6 A Roman historian and senator (d. 117). He had two principal books, *Histories* and *Annals*.
7 Ibid, p. 247.
8 Bacon, *New Organon*, p. 51.
9 Eamon, "From the Secrets of Nature to Public Knowledge", *Reappraisals of Scientific Revolution*, s. 350.
10 Susan Dalton, *Engendering the Republic of Letters: Reconnecting Public and Private Spheres* (Montreal: McGill-Queen's University Press, 2003), 7–8; Dena Goodman, *The Republic of Letters: A Cultural History of the French Enlightenment* (Ithaca-London: Cornell University Press, 1994), 17.

Republic of Letters (*Respublica literaria*) was among the first institutional examples of these solidaristic structures. The only solid existence was the letters. Many scientists in Europe endeavored to build a shared mind by exchanging letters and texts in the seventeenth and eighteenth centuries.

11 This experience is interesting in respect to the connotation for the correspondence tradition in the Islamicate civilization and the schools established through this tradition. Even though correspondence seems to be a daily ordinary phenomenon, it played a significant role for specificity of the tradition of Islamic thought.

12 But following the establishment of French Académie des Sciences, the connection with the French scientist was lost. For the dynamics between the nation-state and modern science, we should analyze the establishment and evolution of these societies.

13 William Eamon, "From the Secrets of Nature to Public Knowledge", *Reappraisals of Scientific Revolution*, p. 343. By the end of the sixteenth century, 15 million prints were done for 35,000 books. See Ian Inkster, "Potentially Global: 'Useful and Reliable Knowledge' and Material Progress in Europe, 1474–1914", *The International History Review*, Vol. 28, No. 2, June 2006 s. 246.

14 Eamon, "From the Secrets of Nature to Public Knowledge", *Reappraisals of Scientific Revolution*, s. 341.

15 Francis Bacon, *The Great Instauration* (Whitefish: Kessinger Publishing, 1996), Preface.

16 Ibid, Preface.

17 Bacon, The New Organon, Book II, s. 107.

18 Ibid, Book I, s. 85.

19 Ibid, Book II, s. 107.

20 Ibid, Book I, s. 81.

21 Hans Jonas, *Practical Use of Theory in Phenomenon of Life* (Illinois: Northwestern University Press, 2001), p. 195.

22 Jonas, p. 196.

23 Jonas, p. 198.

24 For the evolution of the concept see. Eugen Rosenstock-Huessy, *Circulation of Thought*, Vol. 1, Dartmouth College (Lecture Notes) (Hanover: Dartmouth College, 1949), p. 11.

25 G.L.E. Turner, "Eighteenth Century Scientific Instruments and Their Makers", in *The Cambridge History of Science*, Vol. 4, Roy Porter (ed.) (Cambridge: Cambridge University Press, 2008), p. 511.

26 Turner, p. 511.

27 Lissa Roberts, "Mapping Steam Engines and Skill in Eighteenth Century Holland", in *The Mindful Hand: Inquiry and Invention from the Late Renaissance to Early Industrialization*, Lisa Roberts, Simon Schaffer and Peter Dear (ed.) (Leiden: Royal Netherlands Academy of Arts and Sciences, 2007), p. 202.

28 Maria Boss Hall, *Promoting Experimental Learning* (Cambridge: Cambridge University Press, 1991), p. 138.

29 Turner, s. 513.

30 Ibid, p. 515.

31 Inkster, s. 270.

32 Turner, s. 514.

33 Ibid, p. 519.

34 Ibid, p. 521.
35 Ibid, p. 521.
36 Ibid, p. 521.
37 Lissa Roberts, "Science Becomes Electric: Dutch Interaction with the Electrical Machine during the Eighteenth Century", *Isis*, Vol. 90, No. 4, December 1999, s. 700.
38 Turner, s. 522.
39 Turner, s. 522.
40 The first use of microscope began after 1625. Jan Swammerdam (1637–1680), a Dutch microscope producer, observed the future wings of a butterfly within a caterpillar for the first time and took this as interference of the omnipotent God into this insect's anatomy. It took almost 70 years for the usage of microscope in practical matters by Hooke. Even John Locke, a philosopher who was interested in medicine, found microscope as a useless instrument. See Henry, s. 44.
41 Laurance Brockliss, "Science, Universities and Other Public Places", in *The Cambridge History of Science*, Vol. 4. Roy Porter (ed.), Cambridge: Cambridge University Press, 2008, p. 52.
42 Brockliss, s. 53.
43 Ibid, p. 54.
44 Ibid, p. 55.
45 Ibid, p. 55.
46 Ibid, p. 56.
47 Ibid, p. 56. We can count a few examples of this kind: Albrecht von Haller (1708–1777), an expert in medicine and botanic; Friedrich Blumenbach (1752–1841), an orientalist, and Christoph Lichtenberg (1742–1799), a philosopher whom we know from his experiments of electricity.
48 Immanuel Kant, *Conflict of the Faculties*, Mary Gregor (trans.) (New York: Abaris Books, 1979), p. 12.
49 Brockliss, s. 59.
50 Ibid, p. 59.
51 We can count a few examples, such as Jacques Rohault (d. 1672) who introduced Cartesian philosophy in many French towns, John Keill (1671–1721) who began to teach Newton physics and his doctrine at Oxford in 1694, William Jacob van's Gravesande (1688–1742) who was one of the most famous professors spreading Newton's laws and a curriculum developer, and Pieter Musschenbroek (1692–1761) who was the inventor of Leiden jar.
52 Laurance Brockliss, "Science, Universities and Other Public Places", *The Cambridge History of Science*, Vol. 4. Roy Porter (ed.), Cambridge: Cambridge University Press, 2008 s. 58.
53 Brockliss, p. 59.
54 Brockliss, p. 72.
55 *Ecole Polytechnic*, continuing to apply higher education programs, is now working under the Ministry of Defense. It is one of the important institutions of the French Revolution.
56 Brockliss, p. 75.
57 Brockliss, p. 81.
58 Brockliss, p. 81. These booths have continued to exist in modern educational institutions at almost all levels and become spaces where those instructed at school were tested.

6 Solidarity between Homo Faber and Homo Economicus

The most important reason for calling modern science as technoscience was the particular relation between *tekhne* (art, technics) and epistemology. Every item taken as problem in the intellectual history and qualified as theory could also be obtained from the practical ground. Modern science seemed to obtain the success in refined systematization of theory-practice relations.

As a matter of fact, we also face a search for legitimacy through the normal relationship between human beings and nature. Tracing the tracks of this search, Ortega defines human life as an existential project and a desire to realize his/her program or a program to realize.[1] His/her social needs, distinct from natural needs, depend on production of a kind of "supernature" so that s/he can control the nature. Therefore, "a man without technology is not man."[2] Human beings must realize themselves willingly or unwillingly.

In this framework, we can take a phase of the seventeenth-century scientific revolution as a "life realizing" project. All these processes, in the circle of needs constructed basically through certain ontological insecurity are spiral associations of their natural cycles with each other.[3] The modern technique is not constructed only as an "answer list for natural needs" but also as a "world view." Therefore, the modern western human beings cannot exist without their technology. Even this technology reached a characteristic enabling them to resist mechanically and even to survive.[4] This set up became possible when above-mentioned phenomenal analysis tools such as scientific instruments or space joined the circle with transformative roles.

Due to the existence of an absolute connection between human nature and human actions, a human search about his/her nature will ultimately be possible through a relationship with the practice, because of

DOI: 10.4324/9781003275756-7

the fact that all kinds of searches about human beings certainly cover the works in practical areas. Thus, we can say that all *post-human* discussions are, indeed, a human search for human nature.

It is useful to follow prototypes forming social structure in order to understand which things modern science has kept as they were and what it has transformed. In this context, it is important to ask through which crucial points human relations have formed and what kind of contribution these social dynamics have had on the emergence of modern empirical science. For example, the "privileged knowledge system," which has been effective since the fourteenth century, played a significant role in this network of relations and led to the emergence of a new synthesis.

According to Max Weber, "an ideal type (*Idealtypus*) is formed by the one-sided *accentuation* of one or more points of view" according to which "*concrete individual* phenomena ... are arranged into a unified analytical construct." [5] Although Weber states that this is an imaginary construction for idealization, it shows that the individual's (and consequently science's) relations with the reality are subjective. In fact, all civilizations create characteristically fictive personalities—*Idealtypus*. In this sense, it is possible to argue that the modern epistemology created an idiosyncratic *Homo faber*[6] and a *Homo economicus*[7] and brought them to the center. Although the existence of an *Idealtypus* comprising both of these qualities is not a necessary condition, all modernization processes came through the capacity of "interactive contact and solidarity" of these two and through the sustainability of this platform due to its force of attraction. The human mobility that came with population growth contributed to the conduct of this solidarity with the modernity for the first time.

This new interaction pattern is important for the social class struggle emerging in Europe as well as the history of science. This process, triggered by all classes that was affected negatively by the system which was as distinct from "medieval communication network"[8] formed by the appearance of monotheistic religions, evolved and transformed the modern science paradigm into a flexible structure transcending national or feudal borders and making them transmitter. This structure that gained a political assurance by the Westphalian system used the modern science paradigm as an intellectual language in general and Galileo's law of motion in particular.[9]

In connection with the analogy to *Deus faber, Homo faber* presented its acumen and talents, developed its feeling of absolute power, progresses significantly in building the twentieth century. Bergson also defines it as the ability of an intelligence with infinite powers of design and creation to make every kind of tools and tool-making tools.[10] *Homo faber* is a circle seeking absolute physical contact between human beings and the universe and even the one enabling this contact. The modern man is able to reproduce and build himself only through this contact.

Under normal circumstances, the individual aims at consuming less energy in order to acquire a luxurious and comfortable life. Even though this idea of efficiency acts on rational ground, the goal is not to obtain a rational result. The idea of *Homo economicus*, which leaves man alone and brings individual benefits to the center, found spaces at various levels in different philosophical schools. Mandeville's *Fable of the Bees* assumes that public good appears out of personal weaknesses.[11] All these appraisals are reflections of the process of building *Homo economicus. Homo economicus* is exposition of the transformation of the balance between philanthropy and selfishness as two pre-modern exchange models into a "mutual consent system" founded in the twentieth century. However, we should not forget that this transformation could fully happen only after a lot of turmoil.

The new approach evoking discrimination of *artes serviles* coming from the antiquity failed to eliminate oppressions and discriminations to the working class. After the seventeenth century, the problems of population growth in cities and extreme poverty posed significant issues for governments. Various suggestions for their solutions were presented. Locke was among those who proposed solutions for the working class. In his work A *Report to the Board of Trade to the Lords Justices*, in 1697, he states that "the state should make laborers industrious through hard work, whipping and torture, and even should have them face the risk of losing ears." Seeing the working class as "lowly class," Locke argues that upper classes should rule them with a rational ethics.[12]

While the industrial revolution is a prerequisite for production-consumption relations, the existing technoscience cannot be explained solely by the industrial revolution. For the creation of *Homo faber*, laborers and labor played a significant role as much as technicians. *Homo economicus* raised *Homo faber* into an upper level during the process of institutionalization of the capitalism and produced working class. An establishment similar to the unqualified labor force in agrarian societies appeared as an important factor during the industrialization process starting with mining in the seventeenth century. The masses could find a space in the system only in the twentieth century through the widening of the working class and the establishment of social state institutions.

While we can find examples of archetypes of *Homo economicus* and *Homo faber* in old periods, their appearance as systematic structures fostering each other could be made possible only in the modernity.[13] Thus, we can view the modern *tekhne* as a shared product appearing out of solidarity between *Homo faber* and *Homo economicus*. The contribution of human nature to modern epistemological accumulation is a testimony to the fact that this solidarity was profound.

It is one of the salient aspects of modernity that *Homo sapiens* did not remain as a whole, but stood fragmented.[14] According to Scheler, practical knowledge is the default form of knowledge that human beings inherently possess. Due to the fact that human beings are practical creatures, they endeavor to make their future secure through dominating their surroundings and achieving the desired outcomes. Yet, since the relationship between human beings and practice is not natural, human beings have the ability to know and understand nature in respect to their existence. Scheler, who defines philosophy as "the love to join the essence of the whole existence with the human center," states that human anxiety and interest led human beings to transcend the practical to philosophical dimension.[15] The practical deprivations led, in some sense, human beings to seek security, and caused them to enter into a philosophical research. This effort shows the background source of the modern *tekhne* and epistemology.

If we return to the solidarity between *Homo economicus* and *Homo faber*, we can say that contemporary modern epistemological order was founded by acquiring power from a certain point of ontological insecurity. Before the industrial revolution and the invention of steam engine, the feeling of "poverty" and "weakness," and the fading of the idea of a divine savior and also the perception of this solidarity as a new savior were determinants. This solidarity is a dynamic collaboration that is transmitted to generations through social memory, but it needs to be reestablished each time. The fact that a physical association caused a transformation (that we can call chemical in some sense) is closely related to the fact that the Western world where this transformation took place had deprivation at a certain time. Thus, it is possible to argue that it could not yield similar results in another society and geography.

The analyses of spaces and environment focused on laboratory, academies and *Kunstkammer* in which *Homo economicus faber*[16] evolved. The history of science and the arts can reflect episodes of the history of *Homo faber*. Human relations with tools and artistic solutions for daily needs played a significant role in all "life realizations," but its determinative role appeared only after the emergence of *Homo economicus*. While Da Vinci and Galileo were among the most important Western examples of *Homo faber*, Francis Bacon was an example for the synthesis of *Homo economicus faber*. During the period before *Homo economicus*, the dominant behavior model was *Homo reciprocans* (*philanthropic man*)[17] to define those who made tools and used them in scientific researches, in astronomy and alchemy, developed mathematical tools and predominantly prioritized sharing rather than social and occupational competition. The association of *Homo reciprocans* with *Homo faber* in this sense became a foundational element of all classical civilizations.

As the biological and non-biological collection, what we call human being, continues its existence in a fragile structure in the ecosystem, all solidarity models aim at resolving the ontological insecurity problem, thereby protecting human being.

6.1 From *Tekhne* to *Phronetic Tekhne*

All technical processes have a *tekhne* meaning free from its temporal and spatial circumstances and this meaning can be internalized. Therefore, the technical processes produced in the Western world would find a reflection in the rest of the world and another human being would be built. This building process is necessarily in touch with the practical life. Every element that has value in the practical life will rule as factors shaping the process.

As the modern *technoscience* is a special synthesis depending on the ontological security index, its transmission and internalization has become possible under special circumstances. According to Michael Hanby, human being fragmented the universe by what his/her made and produced and these fragments appear both as objects and as subjects seeing and manipulating its own nature as object.[18] Since this interaction holds a structure beyond human control through many technical, bureaucratical and political processes, comfortable and humanistic life is impossible. The otherwise depends on a realistic and sincere understanding of art and this kind of art depends on human rediscovery of oneself by recreating.[19] Hanby asserts that all instruments that human being made for his/her service reformulate oneself and even all technical and political processes have evolved into a structure limiting or precluding human control because the scale of this ontology exceeds human level. So many post-human views are based on similar arguments. The emergence of a more humanistic *tekhne* is impossible as long as the institutions endure under such a domination.[20]

Beside to every *tekhne* has an assumption on human being, it also creates its own ontology on another human level. In addition to internal functional value, *Tekhne* faces secondarily a new dynamism. In this encounter, the "ontological security index" steps in primarily as an external variable and the process of explanation begins again. This explanation foresees a new hierarchy, through which the ontological security is established. It is possible to conceive a *phronetic tekhne* within the framework of physical principles.[21] *Phronetic tekhne* would bring a limit to the insolence that could possibly appear by the emergence of a different communicative or interactive mode as a result of the internal transformation of the established practical order or by an anxiety of sustainability in case of a possible self-destruction of the present order. If the first scenario comes true, the new mode of relation that appeared especially by the globalization would build a *tekhne*, which would be meaningful for humanity, sustainable, absolutely secure, by reordering habits of production and consumption that would transform the existing structures.

Notes

1 Ortega y Gasset, *Man as a Project*, Samuel Moody (trans.), Reading for Philosophical Inquiry: Article Series, http://philosophy.lander.edu/intro/articles/ortega-a.pdf, 2008, s. 6 (10 Aralık 2012).

2 Luis M. Rocha, *Technology*, http://informatics.indiana.edu/rocha/i101/pdfs/i101_lecnotes_v2.pdf, s. 2 (23 Mart 2012).

3 To give a timespan, we can extend the period from the translation activities in the twelfth century until the Industrial Revolution.

4 Spinoza's concept of *conatus* summarizes exactly this situation. *Conatus* is an existential resistance as well as the mechanic. See Gilles Deleuze, *Spinoza: Practical Philosophy*, Robert Hurley (trans.), San Francisco: City Lights Books, 1988, p. 101.

5 Sung Ho Kim, "Max Weber", Stanford Encyclopedia of Philosophy, Metaphysics Research Lab, Stanford University, 2012, http://plato.stanford.edu/entries/weber/#IdeTyp (3 Mart 2013).

6 This is a philosophical concept created by Max Scheler and Hanna Arendt. It means "creative human being" who manages the nature and invents tools to subjugate the nature.

7 Joseph Persky, "Retrospectives: The Ethology of Homo Economicus", *The Journal of Economic Perspectives*, Vol. 9, No. 2, Spring 1995, s. 221–31. *Homo Economicus* was first expressed by J. Stuart Mill first as *Economic Man*.

8 We suggest the concept of "Medieval communication network" in order to signify the religion or religious texts as a factor intersecting or uniting all

pre-modern nations and states (besides to climate, environmental effects and ethnicity. It also signifies the networks and communications creating safe and close contact space and political power through the practical language of religion.

9 Ahmet Davutoğlu, *Alternative Paradigms: The Impact of Islamic and Western Weltanschauungs on Political Theory*, London: University Press of America, 1994, p. 169. The Westphalian System emerged as a national political order at the end of Thirty-Year Wars (1618–1638).

10 Henri Bergson, *Creative Evolution*, Arther Mitchell (trans.), Los Angeles: Indo European Publications, 2010, p. 81–2.

11 Bernard Mandeville, *Fable of the Bees*, London: J. Tonson, 1728, p. 212–3. The idea of *"Private vice, public benefit"* states that weaknesses of each individual turn into public benefit with a social order as well as human ambitions are rational behaviors.

12 E. J. Hundert, "The Making of Homo Faber: John Locke between Ideology and History", *Journal of the History of the Ideas*, Vol. 33, No. 1, 1972, s. 8–22. Cf. John Locke, *A Report to the Board of Trade to the Lords Justices (1697)*, London: Respecting the Relief and Unemployment of the Poor, 1789.

13 The "privileged knowledge system" became the principal source creating and fostering this systematic structure.

14 Max Scheler mentions human insight. *Homo faber* begins the effort to combine the sights of "Animal Rational" and "God's Child", in other words the effort to obtain human being.

15 Zachary Davis and Anthony Steinbock, "Max Scheler", Stanford Encyclopedia of Philosophy, Metaphysics Research Lab, Stanford University, http://plato.stanford.edu/entries/scheler (2 February 2011).

16 Triple names are used for classification of animals in Latin.

17 Samuel Bowles and Herbert Gintis, *Homo reciprocans: A Research Initiative on the Origins, Dimensions, and Policy Implications of Reciprocal Fairness*, http://www.umass.edu/preferen/gintis/homo.pdf, 1997 (15 April 2013).

18 Michael Hanby, "Homo Faber and/or Homo Adorans: On the Place of Human Making in a Sacramental Cosmos", *Communio: International Catholic Review*, No. 38, Summer 2011, p. 223.

19 Hanby, s. 224.

20 Hanby, s. 232–3.

21 The concept of *phronetic tekhne* is used as a reference to the ancient use of *phronesis* to explain technical environment that is true for human being. *Phronesis* was a concept in the classical Greek thought meaning "practical wisdom" and being used to distinguish it from theoretical wisdom (*sophia*). In its widest meaning, it is the capacity to use correct tools to obtain best results for human being or the practical wisdom for a happy life. For example see Aristotle, *Nicomachean Ethics*, 2003, p. 327.

Conclusion

This study trying to show the relationship between the nature of *tekhne* (art, technics) and modern science examines the philosophical concepts related to the subject and at the same time adopts a methodology in associating all elements that became subject of the philosophy of nature with their categories of space, human and concepts. Since it would not be possible to show the mutually comprehensive relation between the "practical" and the "theoretical" with a pure philosophical abstraction, this study has reached the conclusion by taking into consideration the idea of "scientific revolution" shaping the foundational philosophy of modern science and the philosophical and historical background of the institutions particular to the period.

When we apply practical definitions of ethnomethodology to the philosophy of science, the result becomes the fact that the legitimacy of the scientific reality is solidified in practical life and this occur only by an internalization at an ontological level. The success of the instruments developed for facilitating and explaining life depend on the order they can create.

Neo-Platonist and other elements that began to appear with translations from Arabic in the thirteenth century became the constituents of knowledge acquired in a new framework. The most crucial influence coming from Islamic thought was probably that the classical differentiations about the philosophy of nature was based only on assumption and by definition they should be taken with caution, thereby it was a search for "reproducible" and "testable" knowledge. This process became possible through the association of the "quantification," initiated by Ibn Haytham, Ibn Bajja and their followers, to epistemology. This quantification, which also reflected in visual and practical arts, enabled to test the perception of reality through some factors like perspective technique on canvas and also to build it through a rational language. The mechanical arts could also be reproduced and transacted by means of this language.

DOI: 10.4324/9781003275756-8

The local dynamics that could not find a space in the comprehensive scholastic system obtained a competitive power through their practical advantages by benefiting the boons of vernacular languages. This advantage contributed significantly to the nationalization in Europe. "Life realization" emerged as a Baconian project as the mechanical arts moved to the center. Transcending Aristotelianism and scholasticism of the Church institution could be possible only through a new constructing reality by a new language. As a matter of fact, Galileo and other contemporary important figures wrote their works in vernacular languages like Italian, English and French, which testifies their competitive advantage and that the new system focused only on realities. Therefore, they could find a way to escape from the scholastic conceptual framework that maintained its dominant position through Latin.

The scientific revolution or the developments in the sixteenth-century European thought of science and art, contrary to what is assumed, should be defined as results of "imperfect individual" based on ontological insecurity coming from class conflicts rather than as a result of a new capability or leap that emerged only in the West and was absent in the other geographies. In fact, the *tekhne* activities after Renaissance emerged, following the trends in Renaissance, as a result of the transformation of mercantilism as well as of the use of the philosophy of nature for "life realization" purposes in the solidarity of *Homo economicus* and *Homo faber* for survival and have social promotion. The existing lack of ontological security changed its space with a new imitation through this solidarity. Experimental science as an application of the philosophy of nature entered into university curriculum 350 years after the beginning of the processes "privileged knowledge order" and "quantification." The leap from *Artes serviles* to "philosophical apparatus" obtained continuity by using the "ontological security" of the philosophy of nature and became quantified by a new imitation.

We can deduce a few conclusions: First, experimental or mathematical philosophy of nature did not originate from a rational advantage particular to the West that the whole society concur on or approved with a common reason. Second, experimental philosophy of nature functioned as unchanging "continuous rectifier"—largely as a mathematical reality based on mechanical arts—and allowed for a transformation of mentality. Third, this transformation was facilitated by the emergence of the prototype *Homo economicus faber* rather than a Scientific Revolution that emerged by its own. The sustainability of this synthesis now is because of its capacity to return "outcomes" into the epistemological system as "inputs," which necessitates the ongoing solidarity in a natural and economic sense. In addition, this solidarity allowed for the use of colonial sources at optimum level and for power accumulation; thus, both *Homo economicus* and *Homo faber* secured their positions by obtaining advantage in the social competition.

In this case, the mathematical and experimental principles of the philosophy of nature on which the modern *tekhne* entering into a new corner with the Industrial Revolution was based were not novel but they were made of ontological security index reaching a critical accumulation by a natural motivation. Especially, as it is clearly seen in al-Farabi's discussion of the definition of *ilm al-hiyal*, it is quite possible to observe the facts of "proto philosophical apparatus", *scientia experimentalis, true scientia, artes mechanicae* and *Homo faber* coming from the cooperation of a philosophical mind and mechanical hand:

> Ilm al-Hiyal or the Science of Measures is the science on what kind of measures should be taken to show words and discussions on natural bodies and everything whose existence is proven and that these proofs are in agreement to each other. All of these sciences deal with lines, planes, bodies and other things as they are rational and natural objects. When generating these proofs in sensible objects by natural objects and bringing them out by a will was desired, a force was needed to organize their agreements and put them in order. A measure is necessary to remove the inhibitors on bringing natural objects into acceptable conditions. Therefore, the science of measures is the science producing knowledge on how these are generated on natural and sensible objects artificially and actually.[1]

Farabi continues to explain it and states that *ilm al-hiyal* is used in equations, in geometric calculations, in making tools for architecture, astronomy and music, and making optical tools giving correct information on the distant objects. However, what is missing here is *Homo economicus*, which is needed in a particular condition to raise *artes serviles* to the level of "philosophical apparatus."

Note

1 Mehmet Farabi, *İlimlerin Sayımı: İhsa'-ül Ulum*, Prof. Mehmet Ateş (trans.), İstanbul: Milli Eğitim Basımevi, 1986, p. 102–3.

Index